TRADE AGREEMENTS, MULTIFUNCTIONALITY AND EU AGRICULTURE

Trade Agreements, Multifunctionality and EU Agriculture

Edited by
Eleni Kaditi and Johan Swinnen

Centre for European Policy Studies
Brussels

The chapters in this book were initially prepared as working papers by the members of the European Network of Agricultural and Rural Policy Research Institutes (ENARPRI). The working papers were presented at a series of workshops organised by ENARPRI member institutes in their respective countries and at the final ENARPRI conference on EU Agriculture and Trade Agreements held in Brussels on 8 June 2006. (See the list of ENARPRI Working Paper titles at the end of this book.)

The editors gratefully acknowledge the support of all the participants at the ENARPRI workshops and conference and are grateful to the network's advisory committee. Financial support from the European Commission under its 5th Research Framework Programme is also acknowledged.

Unless otherwise indicated, the views expressed are attributable only to the authors in a personal capacity and do not necessarily reflect those of CEPS or any other institution with which the authors are associated.

ISBN 978-92-9079-672-5

Centre for European Policy Studies
Place du Congrès 1, B-1000 Brussels
Tel: 32 (0) 2 229.39.11 Fax: 32 (0) 2 219.41.51
e-mail: info@ceps.be
internet: http://www.ceps.be

Contents

Preface

The problems encountered in the Doha development round and the obstacles created by agricultural policy and trade in these negotiations illustrate the importance of the issues addressed in this volume. This volume is a result of research and exchange activities within the European Network of Agricultural and Rural Policy Research Institutes (ENARPRI).

The ENARPRI network was an attempt to address an institutional gap in the EU. An important part of the policy planning and preparation, as well as the international negotiations in the field of agricultural and rural policy is carried out at the central EU level, i.e. by the European Commission. Yet most of the research capacity in the EU on these issues is in member states, within institutes that are the privileged conduit between the academic community and (national) policy-makers. This situation has important advantages, as these institutes assist the member state governments in preparing their positions on agricultural and rural policy. It also allows the integration of local concerns and specific structural conditions into the analysis.

The absence of a central EU research institute, however, constrains both the policy preparation as well as the decision-making process. Moreover, in certain member states agricultural and rural policy research is dispersed and confined within a limited number of small academic units.

The growing importance of international trade negotiations in the area of agricultural and rural policy reinforces the need for an EU-level research capacity. This point certainly holds for trade issues that are negotiated at the EU level.

The purpose of the ENARPRI network was to change this by bringing together leading (national) institutes and research teams from 13 of the 25 EU member states. ENARPRI institutionalised regular meetings within this network and between researchers and users of the policy research, both inside and outside the EU's institutions.

The ENARPRI network was created in 2003 with financial support from the European Commission under its 5th Research Framework Programme. The network operations lasted for four years and its activities included the organisation of workshops and conferences as well as the publication of various working papers, policy briefs and this volume.

ENARPRI led to an improved exchange of information and policy research insights. It also contributed to the development of tools and methods, and the organisation of integrated research programmes. There were significant positive spillover effects and economies of scale within the network by avoiding overlap in the development of models for quantitative evaluations as well as policy scenarios, and by linking the comparative advantages of various institutes through network collaboration. The activities of the network were coordinated and managed by CEPS.

The central theme of the network over the period 2003–06 was the impact of regional, bilateral, and multilateral trade agreements that the EU has concluded or is negotiating, including those associated with the World Trade Organisation, EU enlargement, the Everything but Arms initiative, the Euro-Mediterranean Association Agreements (EuroMed Agreements) and Mercosur. Most of the agreements are complex in nature and require significant modelling efforts to analyse the effects in sufficient detail in order to make the results useful. The agreements have repercussions not only on efficiency and growth, but also on income and welfare distribution within the EU. To analyse these effects in detail, collaboration among various institutes with specific knowledge of local circumstances and data benefited the overall effort. Furthermore, several of the trade agreements have significant interaction effects, which accentuated the benefits arising from a concerted effort. Within this general theme of looking at trade agreements, the network paid particular and extensive attention to the impact of the EuroMed Agreements/trade proposals and their interrelation with other trade agreements.

Another central theme of the network was the multifunctional model of European agriculture and the sustainable development of rural areas. The impact of the trade agreements on the structure of EU agriculture and the livelihoods of rural areas is especially important, as is the interaction of the trade agreements with EU policies. A significant debate is taking place in the EU about the need to revise some EU agricultural and rural policies to address existing concerns about the sustainability of EU agriculture and rural areas. Obvious questions emerge as to whether some of the proposed policies, such as payments for good farming practices or for agri-environmental purposes, are consistent with some of the trade agreements. Hence, the correlation between EU policies within the multifunctionality and sustainable development framework and trade agreements was an important focus of the network.

This volume presents some of the findings of researchers involved in the ENARPRI network over the last four years. The chapters of this book were initially presented as working papers at a series of workshops organised by ENARPRI member institutes in their respective countries. All the chapters were presented at the final ENARPRI conference on "EU Agriculture and Trade Agreements", organised by CEPS and held in Brussels on 8 June 2006.

I hope readers find the material in this book interesting and relevant. As explained above, however, these written papers represent only (a minor) part of the ENARPRI results, as the institutional development of cooperation and exchange alongside the policy dialogue at the member state and EU level have been the major results.

Let me end by acknowledging the support of all the participants at the ENARPRI workshops and conference. I would also like to thank the network's advisory committee, which was composed of Professor Giovanni Anania (University of Calabria), Professor Jean-Christophe Bureau (University of Paris), Dr Elizabeth Guttenstein (World Wildlife Fund) and Dr Marina Mastrostefano (European Commission).

Johan Swinnen
ENARPRI Coordinator
Senior Research Fellow, CEPS
Professor, Katholieke Universiteit Leuven
Brussels, December 2006

Introduction

Eleni Kaditi

The EU's common agricultural policy (CAP) has a long history of reforming and adapting various policy instruments to protect its agri-food markets from competition from non-EU farmers and food processors. These instruments include tariffs and safeguard measures, domestic and export subsidies, as well as health and safety standards and other types of regulatory barriers. Restrictions on their use arise from the commitments of various multilateral trade negotiations, primarily from the EU's membership of the World Trade Organisation (WTO). Moreover, the EU has made a policy choice to provide support for its farmers based on the objective of ensuring a sustainable agricultural sector in the Community, including not only economic, but also social and environmental criteria. This objective can be summarised under the heading of 'multifunctionality'. In the multilateral context, rural development policies thus reduce the impact of EU policies on trade and make the instruments of the CAP less trade-distorting while being fully compatible with the WTO obligations.

The present book provides a synthesis of the results of various analyses related to EU agricultural policies, trade agreements and the issue of multifunctionality. The analyses focus on the impact of regional, bilateral and multilateral trade agreements that the EU has concluded or is negotiating, as well as on the interaction between EU policies and trade agreements within the multifunctionality and the sustainable development framework. The book is divided into two parts: the first part includes a series of preparatory and empirical papers on issues related to EU agriculture and trade agreements, whereas the second part examines the implications of trade policy for the multifunctionality of EU agriculture, with reference to a series of country case studies.

In particular, chapter 1 by Salamon et al. provides an overview of multilateral trade agreements and the EU's membership of the WTO, with an emphasis on the impact of resultant commitments for EU agriculture. The preferential trade agreements of the EU that affect its agriculture are also illustrated, along with the case of free trade agreements. In addition the analysis covers all the potential changes in the use of various EU trade policy instruments affected by its trade agreements, and the interactions between the latter and the EU's domestic policy instruments. Finally, some of the specific requirements needed to model the effects of changes to trade policies are discussed as a precursor to the impact analyses included in the empirical chapters in the first part of this book.

Chapter 2, by Kuiper, outlines a number of issues related to the Euro-Mediterranean Partnership (EMP). The analysis starts with a brief presentation of the current situation in the negotiations related to the EMP, followed by a description of the accompanying financial support and the Euro-Mediterranean trade agreements. The discussion continues with the identification of three main factors that influence the impact of the EMP on the economic growth of the Mediterranean partner countries (MPCs). These are the amount of liberalisation achieved by the EMP and the factors affecting economic growth as well as trade liberalisation in this region. As far as agriculture is concerned, the author argues that the current implementation of the EMP is unlikely to have a significant result in the reduction of agricultural trade protection for either the EU or the MPCs. To promote stability and economic growth in the MPCs, the structural features of their economies have to be considered, together with a coherent EU trade and foreign policy and the promotion of South–South integration.

In chapter 3, authors Jensen and Yu provide the first empirical analysis, examining the interactions of outcomes from the WTO negotiations with domestic policy reforms. The simulation set-up considers a very detailed implementation of the mid-term review of the CAP. The authors argue that it is feasible for the EU to undertake the largest cuts to its final bound aggregate measurement of support and total trade-distorting domestic support (70 and 75%, respectively). Based on this, a tiered reduction formula for other WTO members is proposed. In addition, CAP-related modelling is discussed, with consideration given to relevant WTO features, especially the decoupling of its amber and blue box programmes. The key findings of the numerical simulations are as follows. First, a structural adjustment in EU agriculture and food production is expected, with the outputs of wheat, oilseeds, plant fibres and bovine meats dropping significantly. Second, the EU's net export position in these products is forecasted to deteriorate in response to the reform. Yet, the overall size of the EU's agricultural production and trade would remain nearly unchanged. Third, despite substantial allocative efficiency gains accruing to the EU from the CAP reform, the impact on its terms of trade is found to be quite small. Lastly, the welfare and trade expansion effects on the rest of world are expected to be rather limited on aggregate, as compared with what can be realised from reforms to market access.

Subsequently, chapter 4 by Brockmeier et al. contains another empirical analysis, investigating the effects of possible WTO negotiation outcomes on the EU and third countries. An extended version of the Global Trade Analysis Project (GTAP) model is used to first project a base run that includes Agenda 2000, the EU's enlargement, the Everything but Arms agreement and the mid-term review of the CAP. The policy simulation also includes the WTO negotiations. For this latter aspect, the simulations differentiate among four proposals that have been submitted by the EU, the US, the G-10 and the G-20 in advance of the WTO's Ministerial Conference in Hong Kong in October 2005. An adequate tariff-line representation is provided, which takes applied and bound rates into account. The authors conclude that highly protected sectors would experience severe negative changes to their trade balances as a result of all four proposals, but particularly under the application of the US proposal. It is also evident that the highly protected beef and milk sectors of the EU would be especially affected.

The last chapter in this part of the book, chapter 5, includes an empirical analysis of the impact of the EMP on trade flows and economic development of the MPCs. Kuiper and van Tongeren focus their study on two countries in particular, Morocco and Tunisia. The authors investigate the effects of both the current Euro-Mediterranean Association Agreements (limited to trade in manufactured goods) and a scenario involving full trade liberalisation, which extends to the agricultural sector. In terms of domestic policies, the authors consider the cases of non-replacement and full replacement of tariff revenues by a consumption tax. Using the GTAP model, their results show a strong potential for increasing earnings and employment in both countries. These gains depend on whether tariff revenues need to be redeemed through domestic taxes or otherwise. Moreover, it is concluded that EU member states are not significantly affected even when agricultural trade is fully liberalised. As a result, the Euro-Mediterranean Agreements could be aligned with domestic reforms in the respective countries. This approach could reap the full potential benefits of the agreements for the MPCs.

The second part of this book starts with an introductory chapter by Dwyer and Guyomard on issues related to multilateral trade agreements and multifunctionality. In chapter 6, Dwyer and Guyomard discuss all the practical difficulties of bringing together models that examine the economic impact of trade policy reforms and models that can measure environmental or multifunctionality indicators. A brief overview of all the country case studies included in the proceeding chapters is also provided, together with an analysis of a varying number of indicators of multifunctionality. In particular, the case studies cover four different EU member states, i.e. Finland, the Czech Republic, Ireland and Greece. As explained below, all the chapters develop different approaches for analysing the effects of alternative policy scenarios on the multifunctional role of agriculture for the respective countries.

In chapter 7, by Lehtonen et al., a sector-model approach is used to predict and compare the multifunctionality effects of various agricultural policy reforms in Finland. Using the DREMFIA model, the key finding of their analysis is that reform of the CAP is not likely to result in any drastic decline of agricultural production in Finland. The cropped area is predicted to decline significantly as more land is put into green set-aside areas when agricultural support payments are decoupled from production. The agricultural labour force is likely to decline substantially irrespective of agricultural policy. This study further concludes that the credibility of the production economics and biological relationships of the economic model determine the validity of the results for the numerous indicators examined. Finally, it is argued that the economic logic of microeconomic simulation models provides a consistent assessment of the many aspects of multifunctionality.

The case study in chapter 8, by Doucha and Foltýn, assesses current trends in the multifunctionality of Czech agriculture. The authors start their analysis with a brief review of the Czech agricultural sector over the last 15 years. They then apply a non-linear optimising model (FARMA-4) to simulate the effects of different policy scenarios up to 2010 on the selected set of indicators of multifunctionality for eight farm categories. Under all the scenarios considered, they conclude that there is a tendency towards a more intensive level of production with less labour input, particularly by the profit-oriented farms that now prevail in the Czech Republic.

Chapter 9 includes an analysis of the relationship between a potential reform of agricultural trade policy under a WTO agreement and emissions from the Irish agricultural sector. Donnellan and Hanrahan focus their study on the impact of agricultural production levels and practices on the level of greenhouse gas (GHG) and ammonia emissions from this sector. A FAPRI-Ireland model is used to examine the issues under consideration, for which a brief description is given followed by a presentation of the two scenarios used in their projections. The authors conclude that over the next 10 years emissions of GHG and ammonia from Irish agriculture are likely to decline relative to existing levels. Potential WTO trade reforms are also expected to lead to moderate reductions in emissions of GHG and ammonia by 2015.

The last case study is presented in chapter 10, which models the relationship between trade policy reforms and the multifunctionality of agriculture in Greece. Psaltopoulos and Balamou use national and regional social accounting matrices to examine changes in the farm and non-farm sectors at the respective levels in response to policy changes. Having presented the model and methodology used in this study, the authors then specify alternative policy scenarios and estimate their effects. Their overall conclusion is that the impact of policy reforms upon multifunctionality indicators is rather mixed. Under a status quo scenario, a significant decline in agricultural employment is projected at the national and regional levels. In general, it is argued that increased pollution emissions are generated, which overcome the environmental benefits deriving from farming activity contraction, owing to the implementation of specific pillar II policy instruments. The effects of scenarios modelling full decoupling, the elimination of export subsidies and the reduction of decoupled income support under pillar I are shown to be rather worrying for most categories of projections.

Finally, Alexopoulos et al. offer a synthesis of the case study findings in chapter 11. This last chapter in the book focuses on the relevance of key conclusions to both the understanding of trade and multifunctionality interrelations and the domestic and international policy processes. Their analysis closes by specifying issues that call for further consideration and analysis.

Part I

Trade Agreements
and EU Agriculture

1. Key issues surrounding general trade agreements

Petra Salamon, Crescenzo dell'Aquila, Ellen Huan-Niemi, Hans Jensen, Marianne Kurzweil, Oliver von Ledebur and Jyrki Niemi

Introduction

In the current discussion on trade liberalisation, there is a focus on the multilateral negotiations of the Doha Development Agenda (DDA) put forward by the World Trade Organisation (WTO). Yet at the same time, the EU – like other major trading partners and particularly the US – has been concluding a wide variety of free trade agreements (FTAs) and preferential trade agreements (PTAs),[1] whose features are bound by the WTO pacts and, in the case of agriculture in the EU, by the common agricultural policy (CAP). Trade agreements, bilateral or multilateral treaties and other enforceable compacts committing two or more nations to specified terms of commerce usually involve mutually beneficial concessions. These agreements are meant to progressively dismantle trade protections with countries involved in EU enlargement or in the European integration process, and to maintain and deepen economic and political relations with the former colonies of member states. They also aim at facilitating trade with many developing and emerging economies that are seeking to amplify their integration into the world economy and improve trade relationships with the EU (Kurzweil et al., 2003; dell'Aquila et al., 2003).

---

[1] The abbreviation 'TAs' is used in this chapter to refer to customs unions or free trade areas (which imply full tariff liberalisation among members), or other arrangements with limited tariff preferences.

Even though these trade agreements were often developed outside of the WTO framework, the General Agreement on Tariffs and Trade (GATT) and the WTO as its successor provide an umbrella for all other agreements. In this context, specific provisions guarantee the compatibility of regional TAs with the most-favoured nation (MFN) clause, namely to allow for a subset of WTO member countries to increase mutual trade through discriminatory measures[2] under certain conditions. Under the GATT's Art. XXIV (1947), WTO members can form trading blocks discriminating against non-members if the agreements: i) involve free trade (i.e. the elimination of all tariffs) within the block for substantially all products; and ii) ensure that there is no increase in external trade barriers against non-member countries. As with the customs union formed by the European Economic Community (EEC), the various association agreements the EEC concludes with candidate and non-candidate countries and the new partnership agreements with African, Caribbean and Pacific (ACP) countries are done so under the GATT's Art. XXIV. This article is the only mechanism through which developed countries can also be recipients of trade preferences in trade agreements. To monitor the Janus-faced regional TAs, the WTO General Council created the Regional Trade Agreements Committee in 1996, to assess whether or not the TAs are consistent with the WTO agreements.

The next section gives a brief overview of multilateral TAs along with FTAs and PTAs, highlighting their political dimensions. It introduces some of the specific requirements for the impact analyses that are addressed in later sections. Specific regulations are occasionally illustrated by examples from the EU sugar regime to emphasise certain effects of international trade policies.[3]

---

[2] The MFN principle, elaborated in Art. I of the WTO, provides for each WTO member to grant to all members the same advantage, privilege and favour that it grants to any other country. TAs, providing for lower tariffs on goods produced in the member countries than on goods produced outside, would violate the non-discriminatory approach stated by Art. I. Where exemptions are allowed, these are regulated by Art. 24.

[3] Further details on preferential TAs and sugar trade can be found in Huan-Niemi & Niemi (2003).

1.1 The EU and multilateral trade agreements of the WTO

1.1.1 General overview of the WTO

As an organisation, the WTO aims at liberalising trade. It provides a forum for governments to negotiate trade issues and settle trade disputes, and operates a system of trade rules. The WTO was established on 1 January 1995 as the successor to the GATT (1947) agreement. The unofficial international organisation that provided rules and oversaw the international trading system from 1948 until the advent of the WTO was also called GATT, after the agreement itself. Over time, the GATT evolved through rounds of negotiations on trade liberalisation (Table 1.1). Starting in 1986 and ending in 1994, the last GATT round (Uruguay round) led to the creation of the WTO in 1995 (WTO, 2006).

Table 1.1 Overview of the GATT Agreements

Place or name	Year	Main subjects	Member countries
Geneva	1947	Tariffs	23
Annecy	1949	Tariffs	13
Torquay	1951	Tariffs	38
Geneva	1956	Tariffs	26
Geneva (Dillon round)	1960–61	Tariffs	26
Geneva (Kennedy round)	1964–67	Tariffs and anti-dumping measures	62
Geneva (Tokyo round)	1973–79	Tariffs, non-tariff measures, 'framework' agreements	102
Geneva (Uruguay round)	1986–94	Tariffs, non-tariff measures, rules, services, intellectual property, dispute settlement, textiles, agriculture, creation of the WTO, etc.	123

Source: WTO (2006).

Along with other trade concerns, the WTO deals with trade in goods, services and intellectual property. In this respect, three different basic agreements are relevant. While the GATT governs trade in goods, the General Agreement on Trade in Services (GATS) deals with trade in services. The Agreement on Trade-Related Aspects of Intellectual Property Rights (TRIPS) covers inventions, designs and creations.

A number of fundamental principles serve as the foundation of the multilateral trading system:

- **Non-discrimination based on the concepts of MFN and national treatment** implies that countries may not discriminate among their trading partners;[4] furthermore, imported and locally produced items should be treated equally **once they have entered** the domestic market.[5]

- The **lowering of trade barriers by negotiation** aims at reducing customs duties (or tariffs) and the abolition of import bans or quotas that restrict quantities selectively. In addition, non-tariff barriers are to be reduced. The WTO allows countries to liberalise their trade gradually with longer phasing-in periods for developing countries.

- The **binding nature of countries' commitments to market access** provides ceilings on customs tariff rates (**bound rates**). Although countries[6] may tax imports at rates that are lower than the bound rates (**applied rates**), such adjustments are only possible after negotiations with trading partners. The use of tariff-rate quotas (TRQs) and other import-limiting measures is also discouraged, and countries' trading rules are supposed to be **publicly transparent** as evaluated by the Trade Policy Review Mechanism.

---

[4] If a member grants a special favour (such as a lower customs duty rate for one of their products) to one country, this has to be done for all other WTO members. But in each agreement this principle is handled in a slightly different way. Some exceptions exist concerning goods: countries can set up an FTA that applies only to trade within the group; they can also give developing countries special access to their markets or raise barriers against products traded unfairly by a specific country. Exceptions in the trade of services also are allowed.

[5] Therefore, charging customs duty on an import is not a violation of national treatment, even if locally-produced products are not charged an equivalent tax.

[6] For the most part this refers to developing countries.

- **Fair competition** is supposed to be promoted through rules for open and undistorted competition (based on the principle of non-discrimination), which govern issues such as 'dumping' (the export of products at prices that are below cost to gain market share) and the enactment of countermeasures for charging additional import duties to compensate for losses accrued.

- **Development and economic reform** prompts special assistance and trade concessions for developing countries.

From the perspective of the WTO, regional TAs can actually support the multilateral trading system. Issues such as services, intellectual property, environmental standards, investments and competition policies originally arose in the contexts of the regional TAs and were later brought up as topics within the WTO. On the other hand, regional preferential TAs might violate the trade interests of third countries by encroaching on the WTO's principle of equal treatment for all trading partners (MFN) as the case of sugar shows. Here the EU policy on sugar trade has deviated widely from the WTO's non-discrimination principle, given that it applies different policies to different regions and trading blocs (Huan-Niemi & Niemi, 2003, p. 3).

The two largest agreements on goods and services comprise broad principles, extra agreements and annexes dealing with the special requirements, as well as detailed and lengthy schedules of commitments made by individual countries providing access to their markets. For the GATT, these take the form of binding commitments on tariffs for goods in general and combinations of tariffs and quotas for some agricultural products. Underpinning these agreements are the procedures for dispute settlement and the trade policy reviews (see Table 1.2). Important achievements of the Uruguay round were the commitments to 'bind' customs duty rates on the imports of goods, with individual countries listing their legally binding commitments in schedules annexed to the Marrakesh Protocol to the GATT in 1994. Some particularities concerning agriculture and agricultural goods are discussed below.

Table 1.2 Areas covered by the WTO

Umbrella	Agreement establishing the WTO		
	Goods	**Services**	**Intellectual property**
Basic principles	GATT	GATS	TRIPS
Additional details	Other goods agreements and annexes	Services annexes	–
Market access commitments	Countries' schedules of commitments	Countries' schedules of commitments (and MFN exemptions)	–
Dispute settlement	Dispute settlement		
Transparency	Trade policy reviews		

Source: WTO (2006).

1.1.2 Achievements in agricultural trade

During the Uruguay round, tariffs on all agricultural products were bound. Although the original GATT did apply to agricultural trade, it contained loopholes allowing countries to use some non-tariff measures such as import quotas and to subsidise agriculture. The aim of the Uruguay round was to convert all import restrictions into tariffs, a process known as 'tariffication', replacing other restrictions by tariffs giving the same level of protection. The new rules of the Agreement on Agriculture (AoA) and its commitments applied to

- **market access** and various trade restrictions confronting imports;

- **export subsidies** and other methods used to make exports artificially competitive; and

- **domestic support** based on subsidies and other programmes, including those that raise or guarantee farm-gate prices and farmers' incomes.

Provisions concerning market access included the tariffication package with reduced tariffs (Table 1.3), but also aimed at guaranteeing some new access quantities with non-prohibitive duty rates, which was to be achieved by TRQs. For products ruled by tariffication, governments were allowed to take special emergency actions (special safeguards) in order to prevent swiftly falling prices. In the case of sugar, the special safeguard provisions have remained in constant operation since 1995, because of the low prices for sugar on the world market. The fixed standard tariffs and the additional import duties under the safeguard measures have made the import of non-preferential sugar uneconomic in comparison with the price of EU quota sugar in the internal market (Huan-Niemi & Niemi, 2003, p. 4). Under the AoA, countries also used special treatment provisions to restrict the imports of particularly sensitive products (mainly rice) during the implementation period.

Table 1.3 The reductions in agricultural subsidies and protection agreed in the Uruguay round (in %)

	Developed countries 6 years: 1995–2000	Developing countries 10 years: 1995–2004
Tariffs		
Average cut for all agricultural products	–36	–24
Minimum cut per product	–15	–10
Domestic support		
Total aggregate measure of support cuts for the sector (base period: 1986–88)	–20	–13
Exports		
Value of subsidies	–36	–24
Subsidised quantities (base period: 1986–90)	–21	–14

Source: WTO (2006).

In principle, the AoA prohibits export subsidies unless these are specified in a member's lists of commitments. Least-developed countries (LDCs) are exempted. When listed, the agreement requires cutting both the amount of money spent on export subsidies and the quantities that receive subsidies. Taking averages for 1986–90 as the base level, developed countries agreed to cut their values of export subsidies by 36% over six years starting in 1995 and to reduce quantities by 21%. Implementation periods are longer and rates are lower for developing countries.[7] The required reductions did not cause any difficulties in the case of sugar for the EU at the beginning of the implementation period, but towards the end export subsidy commitments became very binding and a 'temporary cut' of 498,800 tonnes in the total A- and B-sugar quotas was needed (Huan-Niemi & Niemi, 2003, p. 3).

The agreement does allow governments to support their rural economies, but preferably through policies that cause less trade distortion. Domestic policies affecting production and trade have had to be cut back. These domestic support measures, which are known as the total aggregate measurement of support (total AMS), were calculated for the agricultural sector and subjected to reductions (except for those operating in the LDCs). Three categories of domestic support were set up. One includes all distorting market-price support measures (with some exceptions), known as the 'amber box', which is defined in Art. 6 of the AoA as all domestic supports except those in the 'blue' and 'green boxes' (as discussed below). Up to 5% of agricultural production in developed countries (and 10% in developing countries) was exempted from cuts (*de minimis* minimal supports), whereas larger subsidies were subjected to cuts. The necessary reduction commitments are expressed in terms of total AMS, comprising all support for specified products as well as non-product-specific support in one overall figure defined in Art. 1 and Annexes 3 and 4 of the AoA.

Other measures with only minimal impact on trade fall into the green box (Annex 2 of the AoA) and can be used freely. Qualified measures must be government funded and do not involve price support. For the most part, this area is where one can find programmes that are not targeted at products, but include decoupled direct income support and programmes for environmental and regional development. If these measures comply with the criteria set out in Annex 2, green box supports are not limited.

Other direct payments coupled to production belong to the blue box, where conditions are applied to reduce distortion. Support normally found in the amber box is shifted to the blue box if the recipient is required to limit production (para. 5 of Art. 6). Currently, there are no limits on spending for programmes in the blue box.

---

[7] For developing countries, these figures are respectively 24% and 14% over a 10-year period.

Again using the example of sugar, the EU was not specifically required to cut its internal price support for sugar under the Uruguay Round Agreement on Agriculture (URAA), because domestic support is measured as the AMS, aggregated across all commodities and policy instruments. Subsequently, the total reduction of 20% over a period of six years for domestic support commitments refers to the total levels of support, but not to individual commodities. Overall, the sector-wide domestic support for sugar has been high compared with other agricultural commodities in the EU, because of the high intervention price for sugar before the reform of the EU sugar regime in 2006 (Huan-Niemi & Niemi, 2003, p. 4).

Creating binding tariffs and applying them equally are the general rule, but the WTO also allows exceptions for taking action against dumping (selling at an unfairly low price) and subsidies. Special 'countervailing' duties are granted to offset the subsidies and emergency measures can be undertaken to temporarily limit imports, designed to 'safeguard' domestic industries. A number of agreements deal with various bureaucratic or legal issues that could involve hindrances to trade such as import licensing, rules for the valuation of goods at customs, further checks on imports, rules of origin, investment measures and licensing procedures.

1.1.3   The Doha Development Agenda

In November 2001, the Fourth Ministerial Conference in Doha launched a new set of negotiations as part of its development agenda. The DDA involves a range of subjects, including those on agriculture and services, on which talks had already begun in early 2000. It also focuses on problems developing countries face under the current WTO agreements. During negotiations, the mandate was refined by work at Cancún in 2003, Geneva in 2004 and Hong Kong in 2005. The negotiations under the mandate of the Doha Declaration were originally supposed to end by 1 January 2005. Governments were to commit themselves to a substantial increase in market access, a possible phasing out of export subsidies and a cut of trade-distorting domestic support. Special and differential treatment for developing countries was made an integral part of the negotiations. The governments also agreed to include non-trade concerns such as those on environmental protection, food security and rural development. At the time of writing, the main features of the negotiations on agriculture could be characterised as below (WTO, 2006).

There are to be three reduction bands in the final bound AMS and in the overall cut in trade-distorting domestic support, with higher linear cuts in the higher bands. The member with the highest level of permitted support will be in the top band and the two members with the second and third highest levels of support will be in the middle band. All others, including developing countries, will be in the bottom band. Overall reductions in trade-distorting domestic support will be required even if the sum of the reductions in the final bound total AMS, *de minimis* and the blue box payments results in a lower figure than the overall reduction required. The EU is in the top band and the US and Japan are in the middle band, but the proposed cuts diverge in most categories. Product-specific AMS may be capped. The blue box will likely be modified with support capped at 2.5% of production value and the green box should be reviewed and clarified.

All forms of export subsidies are to be eliminated by the end of 2013 with front loading, so that substantial cut rates are realised in the beginning of the implementation period. A severe discipline on all export measures with equivalent effects, including export credits, state trading enterprises (STEs) and food aid is to be targeted. Thus far agreement has been reached on the disciplines for short-term export credits, but disagreements remain on how to deal with STEs with monopoly powers and how to ensure that bona fide food aid does not lead to commercial displacement.

Regarding market access, progress has been made on the calculation of tariff equivalents (*ad valorem* equivalents), which build the basis for determining the cut bands. The calculation has been agreed on four tariff tiers, but thresholds still need to be determined. Within the bands, linear cut formulas whose amount is to be negotiated will be applied. Lack of agreement exists with respect to the ceiling on tariffs (between 75% and 100%). Additionally, the treatment of sensitive products must be clarified – at present the proposed figures vary between 1% and 15% of tariff lines. How issues such as TRQs (volume, in-quota tariff and administration), preference erosion and tariff escalation are to be dealt with has yet to be determined. Generally, developing country members will have the possibility of using the Special Safeguard Mechanism (SSG), in which four tariff bands will also be arranged. Furthermore, developing countries are to have the flexibility to individually designate an appropriate number of tariff lines as special products, guided by indicators based on the criteria of food security, livelihood security and rural development. In principle, two-thirds of the industrial country members' scheduled tariff cuts can be devoted to the reductions for developing country members.

Other issues cover subjects such as the treatment of geographical indications, the 'cotton initiative' and other sectoral initiatives, export taxes, tropical products, and special and differential treatment (SDT) for developing countries.

Although it seems as if the disputing parties had reached an agreement in Hong Kong, the actual quantitative outcome of the negotiations, especially those concerning agriculture, remains uncertain. Several options under discussion are presented in the following empirical chapters of the first part of this book, as they form the basis of scenarios carried out and depict the impact of certain proposals.

1.2   EU preferential trade agreements and agriculture

The EU is one of the major players in the global agricultural and food market. European agri-food production as well as trade and trade patterns are governed by the CAP and related trade regimes. Within this overall framework of the WTO, the EU deals with numerous PTAs with third countries (Table 1.4).[8]

In principle, these PTAs and FTAs were originally initiated by different sources. In the Treaty of Rome, provisions were granted for the ACP countries and the overseas countries and territories (OCTs) of the emerging European Community member states. Mutual trade preferences were established in a series of consecutive conventions starting with the Lomé Conventions, which were replaced by the Cotonou Conventions. Whereas previous trade relations had been primarily based on non-reciprocal trade preferences granted by the EU to ACP exports, both sides have agreed to enter into economic integration agreements (new WTO-compatible trading arrangements), progressively remove barriers and enhance cooperation in all areas related to trade. To this end, the Economic Partnership Agreements (EPAs) will be negotiated with the ACPs and applied through a process of regional economic integration. Formal negotiations on the EPAs started in September 2002 and the EPAs will enter into force by 1 January 2008 at the latest. The unilateral trade preferences

---

[8] More details on these agreements can be found in ENARPRI Working Paper No. 1 (Huan-Niemi & Niemi, 2003) and ENARPRI Working Paper No. 3 (Kurzweil, von Ledebur & Salamon, 2003).

will continue to be applied during the interim period 2000–07. Thus the PTAs will be transformed into FTAs.

The OCTs, having had their Association Agreements accepted by the GATT in 1971, have to be regarded separately given that inhabitants of the OCTs are EU citizens. The new association arrangements[9] of the OCTs are designed to promote the economic and social development of the OCTs, to develop economic relations between the OCTs and the EU, to take account of their specific characteristics, and finally, to improve the effectiveness of the financial instruments involved.

The OCTs benefit from preferential access to the EU market. Products originating from the OCTs imported into the EU are not subject to import duties or quantitative restrictions. These arrangements are nonreciprocal; in other words, products originating from the EU are subject to the import duties established by the OCTs.

European integration is another initiative involving the EU and qualifying candidate countries. This process gradually implements a customs union by means of a partial reduction of tariffs or by establishing TRQs (or both), especially in the area of agriculture. Prominent examples in this respect are the Europe Agreements and Association Agreements with Central and Eastern European countries (CEECs), along with Malta, Cyprus and Turkey. Most of the countries involved in such agreements have since become EU member states. These kinds of agreements are now complemented by autonomous trade concessions/agreements with the Balkan countries or those in south-eastern Europe (Albania, Bosnia–Herzegovina, Croatia, the Former Yugoslav Republic of Macedonia, Serbia and Montenegro). They were concluded in 2000 or later as in the case of Albania. The EU has offered the parties to these agreements the possibility of full integration into the EU's structures; thus, as potential accession candidates these countries have been offered tailor-made Stabilisation and Association Agreements. The agreements, over a transitory period, render these trade concessions reciprocal, thereby gradually opening up the markets of the region to EU products.

---

[9] For further details, see European Council, Decision 2001/822/EC of 27 November 2001 on the association of the overseas countries and territories with the European Community, OJ L 314, 11.11.2001 and L 324, 7.12.2001. This Council Decision will remain in force until 2011.

Table 1.4 Overview of the EU's PTAs that affect agriculture

Trade agreement	Countries or regions covered
Europe Agreements	Bulgaria, the Czech Republic, Estonia, Hungary, Latvia, Lithuania, Poland, Romania, Slovakia and Slovenia → EU–CEEC forming EU–25 (1.5.2004) except Bulgaria and Romania
Association Agreements	Cyprus and Malta → EU–CEEC Turkey → EU–RASS
Stabilisation and Association Agreements	Former Yugoslav Republic of Macedonia, Croatia, Albania, Bosnia–Herzegovina, Serbia and Montenegro
Euro–Mediterranean Association Agreements	Israel, Morocco, the Palestinian Authority and Tunisia → EUROMED
Cooperation Agreements (Euro–Mediterranean Association Agreements concluded but not in effect or under negotiation)	Algeria, Egypt, Jordan, Lebanon and Syria → EUROMED
Other FTAs	(Denmark) Faroe Islands, Iceland, Liechtenstein, Norway and Switzerland → EU–EEA South Africa

Table 1.4 cont.

Other customs unions	Andorra and San Marino → EU–OCU
Association of Overseas Countries and Territories	*Anguilla*, Antarctica, Aruba, British Antarctic Territory, British Indian Ocean Territory, British Virgin Islands, Cayman Islands, Falkland Islands, French Polynesia, French Southern and Antarctic Territories, Greenland, *Mayotte*, Montserrat, Netherlands Antilles, New Caledonia, Pitcairn, *Saint Helena, Ascension Island, Tristan da Cunha*, South Georgia and the South Sandwich Islands, *St. Pierre and Miquelon, Turks and Caicos Islands, Wallis and Fortuna Islands* → EU–OCT
EU–African, Caribbean and Pacific (ACP) Partnership	Angola, Antigua and Barbuda, Bahamas, Barbados, Belize, Benin, Botswana, Burkina Faso, Burundi, Cameroon, Cap Verde, Central African Republic, Chad, Comoros, Congo, Cook Islands, Dem. Rep. of Congo, Cote d'Ivoire, Djibouti, Dominica, Dominican Republic, Equatorial Guinea, Eritrea, Ethiopia, Federated States of Micronesia, Fiji, Gabon, Gambia, Ghana, Grenada, Guinea, Guinea–Bissau, Guyana, Haiti, Jamaica, Kenya, Kiribati, Lesotho, Liberia, Madagascar, Malawi, Mali, Marshall Islands, Mauritania, Mauritius, Mozambique, Namibia, Nauru, Niger, Nigeria, Niue Islands, Palau, Papua New Guinea, Rwanda, St. Christopher and Nevis, St. Lucia, St. Vincent and the Grenadines, Samoa, Sao Tome and Principe, Senegal, Seychelles, Sierra Leone, Solomon Islands, Somalia, South Africa, Sudan, Suriname, Swaziland, Tanzania, Togo, Tonga, Trinidad and Tobago, Tuvalu, Uganda, Vanuatu, Zambia and Zimbabwe → EU–ACP

Table 1.4 cont.

Autonomous Trade Measures for the Western Balkans	Albania, Bosnia–Herzegovina, the Federal Republic of Yugoslavia and Kosovo → EU–ATM–Western Balkans
Generalised System of Preferences (GSP) only	*Afghanistan*, Argentina, Armenia, Azerbaijan, Bahrain, *Bangladesh*, Belarus, *Bhutan*, Bolivia, Brazil, Brunei Darussalam, *Cambodia*, Chile, People's Republic of China, Colombia, Costa Rica, Cuba, East Timor, Ecuador, El Salvador, Georgia, Guatemala, Honduras, India, Indonesia, Iran, Iraq, Kazakhstan, Kyrgyzstan, Kuwait, *Lao People's Dem. Rep.*, Libyan Arab Jamahiriya, Malaysia, *Maldives*, Moldova, Mongolia, *Myanmar*, *Nepal*, Nicaragua, Oman, Pakistan, Panama, Paraguay, Peru, Philippines, Qatar, Russian Federation, Saudi Arabia, Sri Lanka, Tajikistan, Thailand, Turkmenistan, Ukraine, the UAE, Uruguay, Uzbekistan, Venezuela, Viet Nam, *Yemen*, American Samoa, Bermuda, Bouvet Island, Cocos Islands, Gibraltar, Guam, Heard and McDonald Islands, Macao, Norfolk Island, Northern Mariana Islands, US Minor Outlying Islands, Tokelau Islands and Virgin Islands (US) → EU–GSP
Everything but Arms (EBA)	The ACP LDCs are *Sudan, Mauritania, Mali, Burkina Faso, Niger, Chad, Cape Verde, Gambia, Guinea–Bissau, Guinea, Sierra Leone, Liberia, Togo, Benin, Central African Republic, Equatorial Guinea, Sao Tomé and Principe, Democratic Republic of Congo, Rwanda, Burundi, Angola, Ethiopia, Eritrea, Djibouti, Somalia, Uganda, Tanzania, Mozambique, Madagascar, Comoros, Zambia, Malawi, Lesotho, Haiti, Solomon Islands, Tuvalu, Kiribati, Vanuatu and Samoa.* The non-ACP LDCs are *Yemen, Afghanistan, Bangladesh, Maldives, Nepal, Bhutan, Myanmar, Laos and Cambodia* → EU–EBA.
Other access	New Zealand, Australia, US and Canada
Cooperative Agreement	MERCOSUR (under negotiation), Chile and Mexico

Note: LDCs are in italics.

Source: Kurzweil, von Ledebur & Salamon (2003).

FTAs were also negotiated with some non-European trading partners to improve relations with developing, emerging and transitional countries as well as better integrate them into the world economy and facilitate trade in both directions. This group comprises the negotiations or signed agreements with MERCOSUR, Chile, Mexico and South Africa.

A broader framework for different preferential regimes is provided by the Generalised System of Preferences (GSP) in conjunction with the Everything but Arms (EBA) initiative. These instruments were established with the dual purpose of facilitating trade and development. But the pertinence of preferential TAs as development measures is still debated. The EBA unilateral trade concession[10] is intended to further improve trading opportunities for the LDCs. All agricultural products are included in the concession, in contrast with the original GSP concession, which focused on manufactured goods. Although LDCs had wide-ranging market access before the concession, afterwards nearly all agricultural products became free from *ad valorem* or specific duties and import quotas. Nevertheless, the full liberalisation of sugar, rice and bananas is being phased in with a transition period. During this period, raw sugar can be exported duty free within the limits of a tariff quota, which will increase each year by 15% from 74,185 tonnes (of white sugar equivalent) in 2001–02 to 197,355 tonnes in 2008–09.) In the case of rice, full liberalisation will be phased in between 1 September 2006 and 1 September 2009 by gradually reducing the EU tariff to zero. In the meantime, a tariff quota has enabled the duty-free access of LDC exports within the limits of 2,517 tonnes (husked-rice equivalent) since 2001–02, rising to 6,696 tonnes in 2008–09. Concerning bananas, after a phasing-in period with gradually reduced tariffs, duty-free access was granted in 2006.

As a result of these processes, the EU has TAs with numerous countries worldwide. The aims and the degree of preferences included in

---

[10] The provisions of the EBA Regulation (European Council Regulation (EC) No. 416/2001 of 28 February 2001) have been incorporated into the GSP Regulation (European Council Regulation (EC) No. 2501/2001). The EBA regulation foresees that the special arrangements for LDCs should be maintained for an unlimited period of time and not be subject to the periodic renewal of the EU's scheme of generalised preferences. Therefore, the date of expiry of European Council Regulation (EC) No. 2501/2001 does not apply to its EBA provisions.

the different TAs may vary a great deal. Within the listed agreements, the EU grants and partly receives numerous preferences that vary according to the date of signature, aim or regional coverage as well as to the kinds of goods targeted. The goal to be achieved by the EU is often more than a provision of economic gains, so targets such as European integration, stabilisation and improved conditions for developing or transitional countries are mentioned. Moreover, while following these political aims, the adherence to European superordinated preferences for domestic goods, especially in the agri-food sector, is apparent for example by the introduction of safeguard provisions and the definition of sensitive products including sugar, beef, dairy products, bananas and other fruits and vegetables. Key features of these agreements are

- They are negotiated **bilaterally**, thus allowing better adjustment during the convergence process, but at the same time implying a greater bargaining power for the EU.[11]

- Nearly all agreements cover **general trade**, whose arrangements are implemented step-wise combined with a gradual removal of tariffs and duties, quantitative import restrictions, export restrictions and export subsidies. To some extent these concessions are then converted into custom unions or 'real' free trade areas, and can be deepened into arrangements for qualifying states to join the EU by applying the *aquis communautaire*.

- Comparably swift adjustments are characteristics of most non-agricultural sectors, but the importance of the CAP implies a much **slower implementation in the agri-food sectors**. Here, tariff reductions are not always assigned in a reciprocal way. Sensitive products of one preference partner are often subjected to lower concessions, as is the case with CAP products such as sugar, beef and veal, dairy products and certain fruits and vegetables.

- Quite often the concessions are granted as **TRQs**, with a fixed import quantity for which a zero tariff or reduced tariff is applied, whereas over-quota import quantities are charged out-of-quota tariffs (normally MFN rates). These quotas can be established for all

---

[11] Exceptions in this respect are the Cotonou Convention and the GSP and EBA protocols.

countries under a certain scheme (e.g. wheat), but are often allocated to individual countries so that unfilled quotas cannot be used by other preferential suppliers (e.g. dairy products). Both systems may be applied to certain goods at the same time (e.g. beef and bananas).

- Tariff reductions and TRQs are always **fixed for specific tariff lines,** and are mostly based on a HS 8-digit level or higher. Some of the tariff preferences are actually temporarily limited to some seasons when production is low in the EU, especially for fruit and vegetables. With regard to the EBA concession, additional access to the EU is provided for 919 agricultural tariff lines at the HS 8-digit level.

- EU duties are often defined as an absolute value in euros per tonne or per 100 kg (specific tariffs). Therefore, the exporter is subject to the effects of varying **exchange rates**.

- The **import licenses** providing the preferences are distributed in various ways, e.g. on a 'first come, first served' basis or given to historical suppliers with quotas for newcomers or implemented through tenders. Different allocation systems may imply different economic outcomes depending on who receives the quota rent.

- From the very beginning, the EU agreements have been equipped with **safeguard clauses** and measures, aimed at protecting the internal market from sudden surges of imports that will affect domestic market prices. In cases where these were not included, they have often been added later on. An example is provided by the new regulation of trade with the OCTs, which now has a safeguard clause. When in 1999 the OCTs tried to circumvent the import quotas imposed, the European Commission applied safeguard measures to prevent the import of sugar and cocoa mixtures from the OCTs.

- A second qualitative restriction is the **rule of origin**, which ensures that a preference is only granted if the product originates in the country to which the preference is provided. The origin of sugar coming from the LDCs could be a concern in the EU.

- Sanitary and phytosanitary (SPS) measures may be set up in the context of the TAs to the extent they protect human, animal and plant health and follow the WTO rules.

- Among other aspects, the agreements refer to **property rights, institutional frameworks,** acceptance of **common standards** and **settings for foreign investments**. Regulations and measurements can prove necessary to generate stable economic growth and income. But under certain circumstances they can turn into obstacles when, for example, common standards prevent some or all imports.

1.3   Policy dimensions

1.3.1   *The policy measures affected*

Looking at the issues and policies covered by the PTAs and FTAs, there is a similar range of measures as those covered by the WTO. All TAs are aimed at lowering trade barriers among their member governments or at least governments try to overcome the most important hindrances to improved trade flows. The main differences in comparison with the WTO are to be seen in the extent to which they affect both trade and domestic support measures.

The multilateral framework of the WTO is better suited for the treatment of domestic policies than PTAs and FTAs. Domestic policies are tested as to whether they are consistent with the WTO's rules or hinder the increase in fair trade, and, based upon that, are also included in the multilateral bindings. Because of EU enlargement, the CAP has been forced to undergo reform for better concordance with WTO rules, through the mid-term review of the CAP. So the impact on the domestic policies of any member state and especially on larger member states is much more pronounced under a multilateral than under a bilateral system.

Although domestic policies are not directly affected by TAs, two important issues for possible policy re-definition remain. First, many existing domestic policies are likely to be challenged owing to their inconsistency with trade regulations, which might be the real core of the negotiations for agricultural liberalisation. Second, lower border protection has an impact on domestic markets and welfare; thus changes in the economic situation of different economic agents can induce adjustments in domestic policies. Naturally, it is to be expected that the effects will increase disproportionally with the economic importance of a trading partner and the negotiated size of the tariff reductions compared with domestic protection. A further question concerns the interaction between domestic policies and trade policies.

Table 1.5 presents an overview of trade policy measures affected by the different trade agreements. Important quantitative considerations include import tariffs, TRQs, export subsidies and the EU entry-price system as well as voluntary export restrictions (VERs).

Table 1.5 Changes in EU trade policy measures owing to EU TAs

	EU–CEECs	EU–RASS	Euro-Med	EU–EEA	EU–ACP	EU–GSP	EU–EBA	EU–Chile/ South Africa	WTO
Tariffs	Yes	Yes	Yes	Yes	Yes	Yes	Yes	Yes	Yes
TRQs	Yes	Yes	Yes	Yes	Yes	Yes	Yes	Yes	Yes
SSG or sensitive products	Yes	Yes	Yes	Yes	Yes	Yes	Yes	Yes	Yes
Import licenses*	Yes	No	No	No	No	No	No	No	No
Entry prices	Yes	Yes	Yes	Yes	Yes	Yes	Yes	Yes	Yes
Export subsidies	Yes	Yes	No	Yes	No	No	Yes	No	Yes
VERs/ non-tariff barriers	No	Yes	Yes	No	No	No	No	No	Yes

* Import licenses are required in general. Trade with the CEECs will eventually become internal trade; therefore no further import licenses will be required.

Source: Authors' compilation.

Import licences are required in all cases, whose distribution can vary by product. Normally, the implementation of a TA does not change this requirement (with the notable exception of EU enlargement). The Agreement on Import Licensing Procedures established by the WTO requires import licensing to be simple, transparent and predictable. Export subsidies are granted to promote exports when world market prices are lower than domestic prices. Not being fixed, export subsidies are evaluated at regular intervals and often adjusted to reflect changes in world market prices, exchange rates, domestic and foreign availabilities, demand, domestic price fluctuations and other changes. Under the WTO agreement, export subsidies were bound multilaterally for the first time and characterised by upper limits to budgetary expenses for export subsidies as well as to subsidised export quantities, both of which were to be reduced. The DDA talks foresee a complete abolition of export subsidies by developed countries by the year 2013. When quotas were banned during the implementation of the WTO, some members obtained similar results by negotiating VERs without technically imposing quotas. Clearly, foreign companies were persuaded to voluntarily restrict the quantities of goods they export to a particular country. The foreign companies agreed because their governments threatened to impose tariffs if the companies did not agree to the VERs. Thus the VERs do not represent a common trade measure,[12] but they are often regarded as a non-tariff barrier to trade. Within the WTO, the technical barriers to trade (TBTs) are regulated by a separate agreement and they will be subject to further discussion in the WTO talks.

Table 1.6 presents the policy linkages between the CAP and the TAs. In general, the policies embedded in the CAP and bilateral TAs are only indirectly related. The CAP measures include

- production quotas (e.g. for sugar, raw milk and starch) and maximum guaranteed quantities (e.g. olive oil, cotton and dried fodder);

- intervention price systems or equivalent regulations (e.g. basic price and target price) in different sectors (e.g. cereals, beef, dairy products, sugar and pork);

- set-aside regulations (e.g. cereals and oilseeds); and

- consumption/processing aids (e.g. butter, skimmed milk powder and cotton).

Usually, these measures are only lightly affected by bilateral TAs, with exceptions for some special cases – e.g. the EBA and the EU sugar sector – in which effects of preferential TAs are considered as very significant as the above-mentioned domestic policy instruments are mostly associated with the amber or blue box of domestic support subject to cuts.

These WTO bindings (and anticipated future cuts) along with the budget issues surrounding the EU's enlargement have induced policy reforms. More specifically, in principle domestic support elements have been shifted from the amber box (market support measures) to the blue box (direct payments) or the green box (income transfers) by the mid-term review of the CAP.

---

[12] Theoretically derived effects of VERs can be found in Bouet (2001) and Wauthy (2002).

Table 1.6 Interactions between the EU's domestic policy measures and EU TAs

	EU-CEECs enlargmt.	EU-RASS	Euro-Med	EU-EEA	EU-ACP	EU-GSP	EU-EBA	EU-Chile/ South Africa	WTO
Production quotas	Yes	(Yes)	(Yes)	(Yes)	(Yes)	(Yes)	(Yes)	No	Yes
Intervention /basic price	Yes	(Yes)	(Yes)	(Yes)	(Yes)	(Yes)	(Yes)	(No)	Yes
Set-aside	Yes	No	No	No	No	No	No	(No)	(Yes)
Consumptio n or processing subsidies	(Yes)	No	No	No	No	No	No	No	(Yes)
Premiums/ single farm payments	Yes	No	No	No	No	No	No	No	Yes
Budget	Yes	(Yes)	(Yes)	(Yes)	(Yes)	(Yes)	(Yes)	(Yes)	Yes

Note: Information in parentheses indicates a very small effect.

Source: Authors' compilation.

As part of the CAP, decoupled support payments represent a further group of policy instruments that may be affected by TAs and vice versa. Whereas the impact of a bilateral TA on the amount of payments and the formal design of direct payments is probably negligible, the effect of payments in the light of a multilateral agreement could be much more significant and distinct. Granting payments enables the EU to lower tariffs and other trade barriers with a limited impact on production. When transfers have been decoupled as scheduled by the mid-term review, the direct support of certain sectors might be lower than would be the case with partly-coupled direct payments; however, production costs will also be reduced by a single farm payment allowing presumably higher production than in a situation where no payments are granted. Thus the EU has greater potential to lower its tariffs when TAs are established. Yet the process of decoupling has actually been prompted by the DDA negotiations, which in all circumstances will require the EU to reduce its protection with respect to import tariffs, export subsidies and domestic support. Under the present system, the shift towards decoupled domestic support would be regarded as the least devastating compared with other protection measures.

All of the instruments discussed above have repercussions on the EU budget, so any change in border measures such as tariffs and export refunds as well as adjustments in the domestic policy instruments, directly affect the EU budget. On the other hand, changes in the EU budget lead to the necessity of policy adjustments, concerning primarily domestic but also indirect trade-policy measures.

1.3.2 Policy representation in models

As stated above, an adequate representation of policy measures in models for trade analysis is needed, covering trade policies as well as domestic agricultural policies. Owing to the wide range of policy instruments carried out by the EU, standard approaches are often used for a stylised representation of policy measures such as the so-called 'tariff equivalents' or 'price wedges'. A tariff equivalent of a policy measure is calculated as the difference between the world market price and the comparable domestic price. A disadvantage of this simple concept is the fact that most of the goods are not homogenous (non-comparable) and undistorted world market prices are seldom obtainable. A common method for deriving tariff equivalents is to use producer support estimates, comprising price distortions, market price supports (transfers from consumers to producers) and transfers from governments to producers. A similar concept exists for consumer subsidy equivalents. The drawbacks are the relevant fluctuations of producer support estimates and consumer support estimates, owing to changes in world market prices, exchange rates and the values of domestic production (van Tongeren & van Meijl, 1999).

Also, the representation of quantitative restrictions by tariff equivalents is often an inadequate approach. Quantitative restrictions are as frequently used in agricultural markets and trade policies as price support instruments. They range from production quotas (e.g. sugar and milk), bounds on intervention buying (e.g. beef and milk), restrictions on livestock production per area, limitations on emissions and overly limited premiums (for most agricultural products), to import quotas, TRQs and limitations on export subsidies. The depiction of these measures is more complicated whenever they pose no real restriction on production or demand. Quotas are then 'non-binding' and the tariff equivalent would be zero or less. Yet policy changes may cause adjustments, leading to a situation in which quotas are restrictive (binding), but the tariff equivalents are still zero or fixed. Model results, in this case, would not detect such a development. Such policy measures can be directly introduced into models, but they require more detailed information.

In all cases, databases are needed that provide information on the policy measures in the regions under observation, as well as data about these instruments. With trade policy measures such as import tariffs, import quotas, TRQs, entry price systems, seasonal import restrictions, export subsidies, restrictions on subsidised exports, export subsidy commitments and non-tariff barriers, a more direct approach is often required.

Table 1.7 gives a brief overview of the approaches to explicitly introducing policy changes in models.

Table 1.7 Requirements for explicit policy representation in modelling TAs

Policy measure	Changing profile or effectiveness owing to PTAs	Modelling of policy instruments	Additional information
Border measures			
Tariffs	Reduced or abolished	Tax rates	Aggregation of tariff lines
TRQs quantities in/out of quota rates	Either newly introduced or increased or abolished (free access)	Complementarity	Aggregation of tariff lines establishing binding and non-binding quotas, fill rates, estimation of rents
Import licenses	Generate or change the distribution of rents among groups	Implementation of rents as income of groups, e.g. importers	Distribution of rents
Export subsidies, limitation on subsidised export quantities	Reduced or abolished	Complementarity	Aggregation of tariff lines establishing binding and non-binding quotas, fill rates, estimation of rents

Table 1.7 cont.

Entry prices (variable levies)	Reduced or abolished	Closure swap	Aggregation of variable levies
VERs	Introduced or abolished	Complementarity	Aggregation of tariff lines establishing binding and non-binding quotas, fill rates, estimation of rents

Domestic policy measures

Production quotas	Reduce rents	Complementarity or upper bounds	Establishing binding and non-binding quotas, fill rates, estimation of rents, distribution of rents
Intervention/ basic price	Changes price wedge	Introduction of additional equations, intervention price and price transmission	Price transmission between the intervention price and market price
Set-aside	Becomes less effective (probably small)	Adjustment of rates	–
Consumption/ processing subsidies	Become less effective (probably small)	Adjustment of rate or closure swap	–
Premiums (animal & hectare), payments (single farm payments)	Affect production levels	Equal or unequal distribution to land or other factors, lump sum	Coupled to factors, full or partial decoupling from factors
Budget	Changes expenditures	Introduction of EU budget	SAM and additional equations, data adjustments to net transfer

Source: Authors' compilation.

As trade policy instruments often comprise quantitative restrictions, the modelling process requires a complementarity approach or, depending on the model used, a fixation of upper bounds. A complementarity approach allows a more detailed representation.

The implementation of these different approaches in models draws upon quantitative country-level information for

- actual, bilateral, applied tariff rates as well as bound rates;

- bilateral TRQs, in-quota tariff rates, actual imports or fill rates, seasonal restrictions and actual imports, price differentiations between in-season and off-season demand, estimations of actual quota rents and distribution of quota rents;

- (bilateral) export subsidies, quantitative value restrictions of export subsidies, restrictions on subsidised exports and fill rates or actual subsidised exports, VERs, estimation of actual quota rents and distribution of quota rents;

- entry prices, 'undistorted world market prices' or variable levies; and

- non-tariff barriers, product standards, rules of origin and SPS measures.

Some data are already available in the new Agricultural Market Access Database (AMAD) established by a joint initiative of the US Department of Agriculture's Economic Research Service, Agriculture and Agrifood Canada, the European Commission, the United Nations' Conference on Trade and Development and the Food and Agriculture Organisation (UNCTAD). When this database is completed, it should contain most of the information on market access commitments on a tariff line basis and their use as needed to model TRQs.

Also, domestic policies are widely displayed in the price-wedge approach. Yet the direct implementation of policy instruments that are subject to limitations may provide better insights, as previously discussed in the case of trade instruments. Intervention price systems need a different approach based on equations on price transmission, while the implementation of the EU budget requires a social accounting matrix and the introduction of additional equations (Brockmeier, 2003).

Again, databases are required providing general information on the domestic policy instruments adopted as well as quantitative data on the policy measures. Particularly, on a country and product basis, data are needed on:

- production quotas, actual production or fill rates, the estimation of actual quota rents and distribution of quota rents;
- subsidies for exports, imports, inputs, factors, outputs and income along with their associated net transfers, restrictions and distribution;
- payments linked to land, animals, quotas and ceilings, decoupled payments as well as their related implementation mechanisms, the degree of decoupling and distribution of transfers among factors;
- limitations on production density, actual production density, distribution of land use and estimations of efficiency changes;
- other transfers;
- set-aside; and
- administered prices, transmission-to-market prices and additional influencing factors (e.g. net exports).

For OECD countries, the components of these data have already been introduced in the OECD's database of producer support estimates and are updated on a yearly basis. The database notably covers data on market price support and the amounts of government transfers to agricultural sectors.

In addition to agri-food and trade policies, other policy measures (particularly those affecting production factors) may have an impact on trade flows. Such policies – not listed in the table, although they can play a non-negligible role – comprise the allocation of land and water rights, the distribution of emission permits, labour policies (minimum wages, union rights and collective labour agreements), migration policies, money supply and interest rates, investment policies, fair trade laws, competition laws, tax policies, monetary policies and so forth. These kinds of instruments can rarely be found implemented in quantitative (trade) models (Brockmeier & Kurzweil, 2003). Nevertheless, their importance increases as the integration process deepens. A further obstacle in quantitative modelling is the fact that such policies have not been established on a long-term basis and policy adjustments occur very regularly.

1.3.3   Further analytical requirements for models posed by EU TAs

The overview in section 1.2 identifies the features of existing EU TAs. Some insights can be derived from these features on additional core elements to be taken into account when modelling the impact of the agreements.

Table 1.8 Requirements for analysing key elements of EU TAs

Requirement	Analytical needs	Modelling
Economy wide	General approach	AGE model
Bilateral preferences	Bilateral trade flows	Multi-region AGE model
Phasing in	Dynamic approach (projections)	Dynamic AGE model, model with projections
Other trading partners	Global approach	Multi-region AGE model
Tariff line representation	Tariff line aggregation	Satellite database of detailed tariff lines
Deep integration process	Adjust for Armington in full enlargement	Introduction of new nests
Product quality (standards)	Armington	Multi-region AGE model
Imperfect competition	Scale economies from fixed costs	Introduce product differentiation on the supply side (need for firm-level data)
Migration	Model structure adjustment	Adjustment of standard equations, introduction of new nest in labour
Unemployment	Model structure adjustment	Adjustment of standard equations
FDI flows	Introduce FDI flows	–

Source: Authors' compilation.

- The EU's bilateral and multilateral agreements cover general trade. This implies a general approach to representing non-agricultural sectors as well, which can greatly affect the market situation and price formation in factor and input markets. A dynamic model approach allowing for gradual implementation would be helpful. Some endogenous and exogenous factor adjustments along the time line may be necessary. The interaction between non-preferential third countries, preferential third countries and the EU (with trade-creating and trade-diverting effects) reveals the need for a global approach.

- Because of their discriminatory nature, TAs should be studied by observing bilateral trade flows, especially when arrangements involve more than one of the EU's trading partners. This requirement has rarely been implemented in computable, general equilibrium modelling.

- Trade flows are often treated with the assumption of homogeneous products. Homogenous goods do not have distinct features and thus the prices of such goods can be pooled. The so-called 'pooled approach' aggregates imports and exports, so all that is needed is a mechanism to balance imports and exports on a market-wide basis and a mechanism to distribute trade shares to the different suppliers as well as to the different demanding agents. The models only explain inter-industry trade and not intra-industry trade. With such an approach modelling can be easier, although it does not address the need to track bilateral trade flows (only a single flow for each actor is mapped, for either imports or exports). A bilateral approach would require a complete set of interactions between each buyer and seller of each commodity and, therefore, substantially more data and parameters.

- Heterogeneous products can be differentiated and there is no need for equalising prices among suppliers. If goods are heterogeneous, then different buyers are willing to pay different prices to obtain the same quantity of a good. Each actor may be both a buyer and a seller at the same time. Product differentiation can be introduced exogenously by assuming that products are differentiated by country of origin. This method was introduced by Armington (1969) by simply assuming that imports of a certain good and domestically produced goods are imperfect substitutes in demand. Combined with a preference function that is separable in domestic and foreign products, it results in (manageable) import functions. This implies an exogenously introduced product differentiation on the demand side. Furthermore, the Armington approach would also allow the implementation of qualitative issues such as product quality and consumer preferences (as represented in, for example, SPS and common standards). In this context, products would not only be distinguished by origin, but also by product quality. In addition, some constraints including small or zero-level trade flows may hinder the impact analysis of any trade liberalisation in the Armington approach. If prohibitively high tariffs induce negligible trade flows, trade liberalisation will only lead to limited changes because of its trade share. Yet there are possibilities to overcome the problem by an exogenous gravity estimation of possible trade share (Kuiper & van Tongeren, 2006).

- Preferences are granted for distinct tariff lines and not for product groups. They are often defined as source-specific and are nearly always limited in quantity. An analysis of these measures would require an implementation of tariff lines or the establishment of a sound method of aggregation.

- Several other aspects that are not directly related to the preferential TAs need to be addressed in modelling work, such as the free movement of labour. Also, institutional frameworks and settings for improving foreign investments require an approach reflecting migration, capital accumulation and factor diversity. Furthermore, changes need to be addressed in factor availability (as a result of changes in land use, irrigation, increased mobility of labour owing to better education, changes in unemployment rates, regulations concerning the influx of labour and adjusted rates of population).

- Different topics within the context of competition also deserve a more detailed analysis. A major item might be the integration of imperfect competition, in cases in which huge international firms or quasi-state trade firms dominate trade and special trade flows.[13] One approach would be to introduce product differentiation endogenously at the firm level on the supply side. Krugman (1991 and 1993), Ethier (1996) and Ethier & Horn (1984) introduced the concept of monopolistic competition to international trade theory. Traditional gains from trade are supplemented by 'non-comparative advantage' gains from trade in the presence of scale economies and imperfect competition. The increase of firms' output leads to positive scale effects and gains from trade in the form of increased variety. Scale economies normally imply that only the support of a limited number of firms is possible. These are consequently imperfectly competitive. Therefore, trade creation supports a larger number of firms, but also a greater level of competition. An advantage of this concept is that it locates product differentiation on the supply side. A disadvantage is the need for firm data to derive elasticities.

---

[13] Additionally, public procurement may also be of importance in this area.

Summary and issues for further work

The EU is implementing various TAs, which are of a bilateral, regional or multilateral nature within the framework of the WTO. Most of the agreements have a complex set-up and differ across commodities, sectors, tools and measures, and also with respect to the timeframe and the regional coverage. Because of this complexity, impact analysis requires significant analytical efforts, moreover because interactions between the different arrangements make it very difficult to overlook the overall economic effects in the countries involved. In order to obtain better insight into the possible effects of these trade arrangements on prices, quantities, incomes and economic welfare, quantitative models are regarded as necessary tools.

Also apparent is the need to analyse issues such as the economy-wide effects of general trade, reflecting bilateral preferences and phasing-in periods as well as trade creation and trade diversion. Owing to the nature of the TAs a more detailed tariff-line representation than that realised by most quantitative models is required to represent most tariff measures, such as tariff cuts, bound and applied tariffs, specific tariffs, sensitive products and entry prices. To a certain extent, this also applies to the implementation of export subsidies. To capture the interaction between trade and domestic policies, detailed domestic measures have to be mapped within the quantitative models. As the focus here is on trade, the three different boxes of domestic support as well as decoupled policy measures such as the introduction of the single farm payment in the EU form a hub for further investigations.

Thus far, only a limited number of additional features have been addressed in ENARPRI's work. Because of actual developments the focus has shifted from regional FTAs and PTAs to multilateral WTO issues, given that (as previously noted) this institution provides the general regulatory framework to which all TAs are subject. With respect to trade policies almost the same requirements and features apply. Still, in the case of WTO regulations the corresponding interactions and the need to adjust domestic policies according to WTO requirements are much stronger. Against this background the impact analysis in the following empirical chapters concentrates particularly on WTO issues.

As many of the above-mentioned requirements have already been captured by the comparative-static, multi-regional, computable general equilibrium model GTAP (Global Trade Analysis Project) it serves as a starting point for further improvement. The standard version provides a representation of the economy including the linkages between farming, agribusiness, industrial and service sectors within a global context. The use of the non-homothetic constant difference of elasticity functional form to handle private household preferences, the explicit treatment of international trade and transport margins and a global banking sector that links global savings and consumption is innovative in the GTAP. Additional features represent perfect competition in all markets as well as profit- and utility-maximising behaviour on the part of producers and consumers. Trade is represented by bilateral trade matrices based on the Armington (1969) assumption. Policy interventions are usually represented through price wedges but can be extended with more elaborate features, especially when it comes to domestic policies. The framework of the standard GTAP model is well documented in the GTAP book.[14] Thus the standard version already covers such needs as the bilateral, global trade flows of general trade, with most PTAs and FTAs already represented in the applied tariffs of GTAP's database version 6. Furthermore, projections for capturing phasing-in can be easily applied.

Special attention is given to the depiction of tariff line representation (at the HS 6-digit level), the transformation of specific tariffs into *ad-valorem* tariff equivalents and the improved modelling of domestic policies, especially single farm payments. As effects may vary according to factor allocation, resource availability and country-specific demands, results are partly displayed by EU member states. In the following empirical chapters, experiments on the effects of WTO regulations on the EU and other countries are presented. In chapter 3, the interactions of WTO negotiation outcomes with domestic policy reforms are investigated. The simulation set-up considers a very detailed implementation of the mid-term review of the CAP. In addition to this CAP-related modelling, relevant WTO features are well represented and considered such as measures in the amber, green and blue boxes. In chapter 4, the effects of possible WTO negotiation outcomes on the EU and third countries are analysed. The effects are projected using an extended GTAP model, which takes into account various additional aspects. With respect to tariffs, the experiment provides an adequate tariff line representation and also considers specific tariffs, accounting for the existence of bound and applied tariffs in both cases. The

---

[14] See Hertel (1997) and the website (http://www.gtap.agecon.purdue.edu/).

extended GTAP structure further allows for consideration of the EU's common budget and the mid-term review of the CAP.

Bibliography

Armington, P.S. (1969), "A Theory of Demand for Products Distinguished by Place of Production", *International Monetary Fund Staff Papers*, Vol. 16, No. 1, pp. 159-78.

Bergstrand, J.H. (1985), "The Gravity Equation in International Trade: Some Microeconomic Foundations and Empirical Evidence", *Review of Economics and Statistics*, Vol. 67, No. 3, pp. 474-81.

Bouet, A. (2001), "Research and development, voluntary export restriction and tariffs", *European Economic Review*, 45, pp. 323-36.

Brockmeier, M. (2003), *Ökonomische Auswirkungen der EU-Osterweiterung auf den Agrar- und Ernährungssektor: Simulationen auf der Basis eines Allgemeinen Gleichgewichtsmodells*. Kiel: Wissenschaftsverl Vauk, IX, 278 p Agrarökonomische Studien 22 [Habilitation].

Brockmeier, M. and M. Kurzweil (2003), "EU-migration in the context of liberalizing agricultural markets", paper prepared for presentation at the American Agricultural Economics Association Annual Meeting, Montreal, Canada, 27-30 July 2003 (retrieved from http://agecon.lib.umn.edu/cgi-bin/pdf_view.pl?paperid=9030&ftype=.pdf).

Brockmeier, M. and P. Salamon (2004), "Handels- und Budgeteffekte der WTO-Agrarverhandlungen in der Doha-Runde: Der revidierte Harbinson-Vorschlag" [Trade and Budget Effects of the Doha Round: the Revised Harbinson Proposal], *Agrarwirtschaft*, Vol. 53, No. 6, pp. 233-51.

Dell'Aquila, C., M.R. Pupo D'Andrea and B.E. Velazquez (2003), *PVS e sicurezza alimentare. Un esame preliminare delle politiche dell'UE e delle implicazioni della liberalizzazione degli scambi*, Working Paper No. 22, INEA Observatory on EU agricultural policies, Rome, November (retrieved from http://www.inea.it/pdf/filespdf.cfm).

Ethier, W. (1996), "Regionalism in a Multilateral World", *Journal of Political Economy*, Vol. 106, No. 6.

Ethier, W. and H. Horn (1984), "A new look at economic integration", in H. Kierzkowski (ed.), *Monopolistic Competition and International Trade*, Oxford: Oxford University Press, pp. 207-09.

Hertel, T.W. (ed.) (1997), *Global Trade Analysis: Modelling and Application*, Cambridge, MA: Cambridge University Press.

Huan-Niemi, E. and J. Niemi (2003), *The Impact of Preferential, Regional and Multilateral Trade Agreements – A Case Study on the Sugar Regime*, ENARPRI Working Paper No. 1, CEPS, Brussels (retrieved from http://www.enarpri.org).

Kuiper, M. and F. van Tongeren (2006), "Using gravity to move Armington – An empirical approach to the small initial trade share problem in general equilibrium models", unpublished manuscript.

Krugman, P. (1991), *Geography and Trade*, Cambridge MA: MIT Press.

———— (1993), "Regionalism versus multilateralism: Analytical notes", in J. de Melo and A. Panagariya (eds) (1995), *New dimensions in regional integration*, Cambridge, MA: Cambridge University Press, pp. 58-89.

Kurzweil, M., O. von Ledebur and P. Salamon (2003), *Review of Trade Agreements and Issues*, ENARPRI Working Paper No. 3, CEPS, Brussels (retrieved from http://www.enarpri.org).

OECD (2001), *Regional Integration: Observed Trade and Other Economic Effects*, Working Paper TD/TC/WP(2001)19/FINAL, OECD, Paris.

Salamon, P., O. von Ledebur and M. Kurzweil (2003), "The Beef Trade Case – Agenda 2000 and the Mercosur–EU Trade Agreement", in *Landbauforschung Völkenrode*, 2/3/2003 (53).

Van Tongeren, F. and H. van Meijl (eds) (1999), *Review of applied models of international trade in agriculture and related resource and environmental modelling*, LEI Report, LEI, The Hague.

Wauthy, X. (2002), *Research and development, voluntary export restriction and tariffs: A comment*, CEREC, Facultés universitaires Saint-Louis and CORE, Brussels (retrieved from http://www.fusl.ac.be/Files/General/CEREC/cahiers/cerec2001_4.pdf).

World Trade Organisation (2002), *Overview, Negotiations on Agriculture, Committee on Agriculture*, WTO, Geneva, 18 December (retrieved from http://www.wto.org.TN/AG/6).

———— (2002), World Trade Organisation Statistics, Table IV.7, Agricultural Products, Trade by Sector, International Trade Statistics 2002 (retrieved from http://www.wto.org).

————— (2003), *First Draft of Modalities for the Further Commitments* (Revision, 18 March 2003), Negotiations on Agriculture, Committee on Agriculture, WTO, Geneva (retrieved from http://www.wto.org. TN/AG/W/1/Rev.1).

————— (2006), *Draft – Possible Modalities on Agriculture by Crawford Falconer (22 June 2006)*, Negotiations on Agriculture, Committee on Agriculture, WTO, Geneva (retrieved from http://www.wto.org/english/tratop_e/dda_e/modalities06_e.htm).

————— (2006), *Understanding the WTO*, 3rd edition, WTO, Geneva (retrieved from http://www.wto.org/english/thewto_e/whatis_e /whatis_e.htm).

2. The Euro-Mediterranean Partnership and the issues at stake

Marijke Kuiper[1]

Introduction

Following a ministerial meeting in Barcelona in 1995, the EU and its Mediterranean partner countries (MPCs) engaged in an ambitious venture of increased economic, political and social cooperation through the Euro-Mediterranean Association Agreements (EuroMed Agreements) and a programme for financial cooperation. Ambitions in terms of economic cooperation were especially high, aiming at a free trade area (FTA) by 2010. This arrangement should create an area of shared prosperity, fostering peace and stability at the southern borders of the EU. The Barcelona process implied a broadening and deepening of trade agreements dating from the 1970s. A meeting marking the 10th anniversary of the Euro-Mediterranean Partnership (EMP) in November 2005 sought to revive efforts to meet the objectives of the Barcelona process.

The current international security situation is one of the driving forces behind the renewed EU interest in the EMP. In this spirit, the guidelines for a common European security strategy call for the promotion of a ring of well-governed countries around the EU, making explicit references to the Mediterranean countries (European Council, 2003, p. 8):

[1] This chapter is based on a joint ENARPRI working paper with C. dell'Aquila (2003). The author would like to thank Henk Kelholt for his able statistical assistance. Financial support through the EU-sponsored ENARPRI-TRADE project is gratefully acknowledged, as is financial support from the Dutch Ministry of Agriculture, Nature Management and Food Quality.

The European Union's interests require a continued engagement with Mediterranean partners, through more effective economic, security and cultural cooperation in the framework of the Barcelona process.

In a similar vein the United States has also intensified its activities in the Mediterranean, as part of a set of presidential initiatives to promote security interests through economic development. Given this interest in enhancing development in the Mediterranean countries, it seems an appropriate time to take stock of the achievements of the EMP.

The aim of this chapter is to provide a general assessment of the impact of the EMP. We start with a short overview of the current state of the negotiations and a description of the two components of the EMP (EuroMed Agreements and financial support). Based on this discussion we identify three main factors influencing the impact of the EMP on economic growth in the MPCs: the amount of liberalisation achieved by the EMP, the factors affecting economic growth in MPCs and those affecting trade liberalisation by the MPCs. We conclude this discussion with an initial assessment of the EMP and by identifying key policy and research issues.

2.1   The Euro-Mediterranean Partnership

At the Barcelona Conference in 1995 the EU deepened its involvement in the Mediterranean area by launching the EMP with 12 MPCs: Algeria, Cyprus, Egypt, Israel, Jordan, Lebanon, Malta, Morocco, the Palestinian Territories, Syria, Tunisia and Turkey. Libya has the status of observer in the EMP; there are no negotiations on a EuroMed Agreement with Libya nor does Libya qualify for financial support. Among the MPCs, Cyprus and Malta have since become fully fledged members of the EU (in 2004), and Turkey has become an EU candidate country. Given Turkey's desire to progress its candidacy towards full EU membership, the EuroMed Agreements between the EU and Turkey extend far beyond the agreements with other MPCs. This chapter focuses on the MPCs that are not seeking EU membership, thus excluding Cyprus, Malta and Turkey.

The EMP is divided into two components: EuroMed Agreements aimed at liberalisation and cooperation in different areas, and financial support provided through the MEDA instrument and the European Investment Bank.

2.1.1 Euro-Mediterranean Association Agreements

The first part of Table 2.1 summarises the current state of the EuroMed Agreements. EuroMed Agreements have been signed with all the countries previously noted except for Syria, with which negotiations are ongoing. Any trade preferences that may be granted may be bound by rules for members of the World Trade Organisation (WTO) (although in practice the WTO does not appear to restrict regional agreements). Thus Table 2.1 also indicates the WTO membership status of the MPCs.

The establishment of an FTA implies South–South integration as well. The EuroMed Agreements, however, are bilateral agreements between the MPCs and the EU, and do not cover South–South liberalisation. Two possible avenues to such liberalisation are the (revived) Greater Arab Free Trade Area (GAFTA) and the Agadir Agreements, which promote regional integration. Participation in these agreements is indicated in the last two columns of Table 2.1.

Table 2.1 MPC participation in EuroMed, GAFTA and Agadir Agreements as well as WTO membership

	EuroMed Agreement		WTO member	GAFTA	Agadir
	Signed	Effective			
Tunisia	1995	1998	1995	1998	2003
Israel	1995	2000	1995	–	–
Morocco	1996	2000	1995	1998	2003
Palestinian Territories	1997	1997[a]	–	–	–
Jordan	1997	2002	2000	1998	2003
Egypt	2001	2004	1995	1998	2003
Algeria	2002	–[b]	Observer	1998	–
Lebanon	2002	2006	Observer	1998	–
Syria	2004	–[c]	–	1998	–

[a] Agreements with the Palestinian Territories are interim agreements.

[b] Ratification of the agreement with Algeria is pending.

[c] The EU's ratification of the agreement with Syria is pending Syria's response to UN Security Council resolutions.

Sources: European Commission website data (retrieved from http://europa.eu.int/comm/external_relations/euromed/index.htm), European Commission (2003) and WTO website data (retrieved from http://www.wto.org).

Although details of the agreements differ across MPCs, the EuroMed Agreements share a number of common themes:

- political dialogue;
- respect for human rights and democracy;
- the establishment of WTO-compatible free trade over a transitional period of up to 12 years;
- provisions relating to intellectual property, services, public procurement, competition rules, state aid and monopolies ('deep integration');
- economic cooperation in a wide range of sectors;
- coordination with regard to social affairs and migration (including re-admission of illegal immigrants); and
- cultural exchange.

This wide coverage of the agreements implies that full implementation would have a considerable impact on the (economic) relations between the EU and MPCs.

Despite an ambitious goal of creating a Mediterranean FTA by 2010,[2] the agreed speed of trade liberalisation varies greatly among manufacturing, agriculture and services. Concessions on manufacturing goods provide for a well-defined, progressive tariff dismantling over a time span of 12 to 16 years. The EuroMed Agreements stipulate the gradual liberalisation of agricultural trade on the basis of traditional trade flows, through periodical revisions of agricultural protocols. The process must be consistent with national agricultural policies and the results of WTO negotiations. For services, the commitment is to abide by the results of multilateral negotiations (on the General Agreement on Trade in Services, GATS), i.e. there are no additional steps taken within the context of the EuroMed Agreements.

The EuroMed Agreements replace the trade agreements from the 1970s. A major difference compared with these earlier agreements is the

---

[2] The reference to the deadline of 2010 might be better understood as an expression of political will that should provide a common discipline to the contracting parties. Actual trade protocols attached to the EuroMed Agreements provide schedules for tariff cuts that are not consistent with such a time target, which is in part owing to the usually very long process of negotiation and ratification of the agreements.

element of reciprocity in the EuroMed Agreements, which (apart from the case of Israel) provides for unilateral concessions by the EU. As a result, a good share of MPC manufacturing exports has had unrestricted access to EU markets since the 1970s. The trade preferences for industrial goods in the EuroMed Agreements are thus in practice quasi-unilateral, favouring the EU.

Reciprocity is not restricted to manufacturing, but also applies to agricultural products. This aspect implies a new commitment by MPCs to introduce preferential measures favouring EU agro-food exporters. Establishing an FTA would thus grant European exporters preferential access to MPC markets, giving them an edge over other competitors. In practice this seems unlikely to materialise. The US is the EU's main competitor in the Mediterranean region. The US already has free trade arrangements with Israel (since 1985) and Jordan (since 2000), and has recently signed one with Morocco. Furthermore, in line with security policies after September 11th, the US launched a Middle East Trade Initiative in May 2003. This initiative mirrors the EMP, seeking to establish bilateral trade arrangements and general support for more outwardly-oriented policies in the Middle East and North Africa. The overall aim is to establish a US–Middle East FTA within a decade, i.e. before 2013. Given the pace at which the US is moving, the establishment of a European FTA with the MPCs would not give European producers an edge, but would only keep them at par with their US competitors.

2.1.2 Financial support through MEDA and the European Investment Bank

The second pillar of the EMP is a new modality for managing financial cooperation, based on an autonomous financial regime with a single budget for the whole Mediterranean area (MEDA).[3] The first MEDA programme replaced the previous five-year bilateral protocols, entailing a three-fold increase in the financial support provided by the EU (€4.6 billion from 1995–99) and a notable enlargement of issues to be tackled.

---

[3] For further details, see Regulation (EC) No. 1488/96 of 23 July 1996 on financial and technical measures to accompany MEDA, OJ L 189, 30.07.1996 and Regulation (EC) No. 2698/2000 of 27 November 2000 amending Regulation 1488/96 (MEDA II), OJ L 311/1, 12.12.2000.

MEDA I was succeeded by MEDA II, which has made €5.35 billion available over the period 2000–06, while the programme has been incorporated in a larger process of restructuring EU cooperation towards development (European Commission, 2001).

MEDA is closely linked to the aim of creating a Mediterranean FTA. Funds are disbursed for improving economic and social policy-making, as well as for direct budgetary support to facilitate structural adjustment programmes (European Commission, 2003). A limited amount of MEDA interventions are meant to support rural development (technical assistance, training, product diversification, environmental and social protection measures).

During the time span covered by MEDA I (1995–99), about 86% of the funds were committed to bilateral cooperation on structural adjustment (15%), economic transition support (30%), socio-economic balance support (29%), the environment (7%), and rural development (5%). Actual MEDA payments, however, were much lower than the commitments, owing to the length of the implementation period for some projects and negotiation issues as well as cumbersome procedures for project approval and management.

In addition to the MEDA funds the European Investment Bank launched a Euro-Mediterranean Investment Facility in October 2002, encompassing all of its lending to the Mediterranean. The funds disbursed by the European Investment Bank promote private sector development. The lending portfolio of €9 billion was supplemented by an additional €8-10 billion (up to 2006) when the Facility was launched.

2.1.3   Summarising the Euro-Mediterranean Partnership

Despite the wide scope of the EuroMed Agreements, the driving force of the EMP is the establishment of a Mediterranean FTA with trade liberalisation taking centre stage. The implicit assumption seems to be that increased trade fosters growth of the MPC economies. The objectives of the EU seem to be both broad and long-term (promoting political stability at its southern borders through increased welfare) as well as more narrow and short-term (unrestricted access for EU manufacturing exports to MPC markets). The absence of well-defined schedules for abolishing protection in agricultural trade also reflects the narrow short-term interests of the EU.

The short-term interests of the MPCs are primarily the additional funds provided through MEDA and the European Investment Bank. Full implementation of the trade agreements implies a major, possibly socially disruptive, restructuring of their economies. This feature of the EuroMed Agreements is clearly acknowledged by the specific allocation of MEDA funds to facilitate the transition. Since the MPCs already have unrestricted access for their manufacturing exports to the EU, their short- and long-term interests are improved access to EU agricultural markets.

The main goal of the EMP, serving the interests of both the EU and the MPCs, is to promote economic growth. The key instrument for achieving this is a Mediterranean FTA. Whether the goal of economic growth will be reached depends first on the amount of liberalisation achieved by the agreements, i.e. the effectiveness of the agreements in liberalising trade. A second determinant of the effectiveness of the EMP is the importance of increased trade relative to other factors limiting economic growth in the MPCs. If current levels of protection play only a minor role, little can be expected from the EMP in terms of promoting economic growth. The remainder of this chapter therefore looks at three sets of issues affecting the impact of the EMP on economic growth in the MPCs: the amount of liberalisation achieved by the EMP, the major factors limiting economic growth in the MPCs and the scope for liberalisation by the MPCs.

2.2   Liberalisation in the Euro-Mediterranean Partnership

Three elements play a role when assessing the achievements of the EMP in liberalising trade: current patterns in protection, current Mediterranean trade flows and the amount of liberalisation achieved by the EMP.

2.2.1   An initial look at current protection patterns

In order to establish the need and scope for liberalisation, an idea of the recent protection levels affecting Mediterranean trade is necessary. Figure 2.1 summarises the trade restrictiveness of MPCs and the EU.

With the exception of Israel, all the MPCs implement repressive trade policies that are also reflected in high mean tariffs. The MPCs do not seem to follow the global trend of reducing trade protection, making it one of the most protective areas in the world (European Commission, 2003).

The numbers in Figure 2.1 refer to trade policies in general, obscuring the considerable protection of the EU on agricultural products. Agricultural trade policies towards the MPCs are governed by a complex system of seasonal preferences for 'sensitive products',[4] with higher tariffs and entry prices for the majority of fresh fruit and vegetables during EU harvesting periods. The MPCs mirror this by protecting 'strategic products'.[5]

Figure 2.1 Trade restrictiveness and mean tariffs

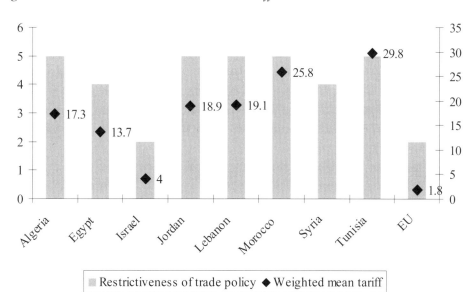

Note: Restrictiveness index: less than 2 = free trade policy, between 4 and 5 is repressed trade policy.

Sources: Heritage Foundation (Gwartney et al., 2003) and World Bank Development Indicators (2002).

---

[4] According to the Commission, 'sensitive' Mediterranean products are tomatoes, olive oil, almonds, oranges, mandarins, lemons, grapes, melons, strawberries, flowers, potatoes, rice and wine (European Commission, 1997).

[5] The definition of 'strategic' products varies among the MPCs, but most of these are staple foods (cereals, meat and dairy products).

In summary, there is ample scope for liberalisation through the EMP. The MPCs could strive for broad trade liberalisation, since their protection levels are well beyond global averages. For the EU, there is plenty of room for simplifying and reducing protection on specific Mediterranean agricultural products.

2.2.2 Current trade flows

Apart from current protection the impact of trade liberalisation in the Mediterranean region depends on the levels of trade. The EU is much larger than the MPCs combined, in terms of both population and the size of the economy. The MPCs are only a minor trading partner for the EU, while the EU is a major trading partner for the MPCs (about 50% of MPC trade is with the EU). The discussion of trade patterns is therefore from the perspective of the MPCs. Since the objective of the EMP is to intensify trade relations in the Mediterranean region, trade data in Tables 2.2 and 2.3 are broken down by region.

Table 2.2 Exports from the MPCs (totals and shares by destination)

	2003–04	Shares (%)				1995	Shares (%)			
	(10⁶ US $)	EU	MPC	US	ROW	(10⁶ US $)	EU	MPC	US	ROW
MPC	*98,260*	*45*	*3*	*22*	*30*	*44,116*	*48*	*3*	*18*	*31*
Algeria	28,347	56	3	22	19	9,357	65	2	17	16
Egypt	7,037	33	9	8	50	3,444	46	11	15	28
Israel	35,201	26	0	37	36	19,047	32	0	30	38
Jordan	3,486	3	12	24	61	1,432	6	8	1	85
Lebanon a)	1,524	9	14	4	73	642	23	13	6	58
Morocco	9,350	75	2	3	20	4,719	62	3	3	32
Syria	5,557	55	10	4	31	n.a.	n.a.	n.a.	n.a.	n.a.
Tunisia	8,520	82	3	1	15	5,475	79	5	1	15

a) For Lebanon data from 2003 and 1997 were used.

Source: International Trade Centre (UNCTAD/WTO) data.

The dominant position of exports to the EU is immediately obvious, accounting for 48% of exports in 1995 and 45% in 2003–04. The (anticipation of the) EMP has not led to an increasing share of exports to the EU; instead there was a slight drop in the overall level of exports to the EU. The aggregate numbers, however, do obscure differences among the MPCs. Both Morocco (13%) and Tunisia (3%) increased their exports to the EU more than to other regions; notably these two MPCs have gone farthest in implementing the EMP. For all other MPCs we find a reduced level of exports to the EU, especially for Lebanon (dropping from 23 to 9%).

Trade among the MPCs does not seem affected by the EMP, remaining stable at 3% for all MPCs. Again, there are differences among the MPCs with both Morocco and Tunisia decreasing their exports to other MPCs, which suggests trade diversion to the EU following the implementation of the EMP. The US is a major player in the region, increasing its export shares in all the MPCs (most strongly in Jordan, jumping from 1 to 24% of its exports). Again Morocco and Tunisia are the exceptions with a low and stable share of exports to the US (the recent FTA between Morocco and the US may change this). The aggregate data in Table 2.2 show that overall exports from the MPCs have increased strongly for all MPCs, and while the relative shares of trade with the EU have declined, trade among the MPCs has remained stable and trade with the US has increased. Data for the two MPCs that have gone farthest with implementing the EMP (Morocco and Tunisia) indicate trade diversion from other MPCs and (mostly) the rest of the world to the EU with a constant share of trade with the US.

Exports only present one side of the trade story; Table 2.3 therefore looks at the pattern in imports. Comparing total imports with total exports reveals a continuing (albeit decreasing) trade deficit for all MPCs. The only exception is Algeria, whose trade surplus in 2003–04 can be attributed to high oil prices. Comparing import shares between the periods 2003–04 and 1995 we find that imports from the EU and US are less significant and there is a small (1%) increase in imports from other MPCs. This situation implies that imports have diversified from the EU and US to other regions. As with exports, Morocco and Tunisia show a different pattern: imports from the EU remain stable while imports from other MPCs are less significant. This outcome again suggests that the EMP promotes trade with the EU.

Based on these aggregated data we can conclude that (anticipation of) the EMP has not led to an increased trade flow in the Mediterranean. Instead, the EU's shares in imports and exports are declining. Meanwhile, the relative share of exports to the US does show an increase. Exceptions to this general trend are Morocco and Tunisia, which have gone farthest in implementing the EMP. This finding suggests that future implementation of the EMP in the other MPCs may intensify trade with the EU, possibly at the expense of trade among the MPCs. Despite a downward trend, the EU's share of imports and exports remains around 50%. This fact implies that liberalisation of trade with the EU would have a considerable impact on MPC economies.

Table 2.3 Imports from the MPCs (totals and shares by destination)

	2003–04	Shares (%)			1995	Shares (%)			
	(10⁶ US $)	EU	MPC	US	ROW	(10⁶ US $)	EU	MPC	US	ROW
MPC	*109,653*	*44*	*3*	*9*	*44*	*78,443*	*51*	*2*	*18*	*29*
Algeria	15,920	55	3	6	36	10,782	59	3	17	21
Egypt	12,112	25	5	11	59	11,739	39	1	15	45
Israel	37,590	40	0	15	44	28,344	52	0	30	18
Jordan	6,898	24	9	7	61	3,696	33	5	1	61
Lebanon[a]	7,167	43	7	6	44	7,438	37	6	6	51
Morocco	16,028	56	2	4	37	8,540	56	3	3	38
Syria	6,080	17	6	5	73	n.a.	n.a.	n.a.	n.a.	n.a.
Tunisia	11,441	70	2	3	25	7,903	71	4	1	24

[a] For Lebanon data from 2003 and 1997 were used.

Source: International Trade Centre (UNCTAD/WTO) data.

The data above do not disaggregate trade in agricultural products. With this being the area in which liberalisation by the EU can be expected to have the most impact, Tables 2.4 and 2.5 present disaggregated agricultural trade data for the MPCs as a whole. Fruit and vegetables are the main agricultural export commodities from the MPCs (33% of total agricultural exports), of which 56% is destined for the EU. This group of commodities is also that for which EU protection is strongest. Cereals are the main agricultural import commodity (29% of agricultural imports), of which 23% originates in the EU and 30% in the US. Across all agricultural commodities, both imported and exported, the EU is the most important trading partner. The liberalisation of agricultural trade in both directions thus has a potentially large impact on trade flows in the Mediterranean.

Table 2.4 Value and destination of agricultural exports by MPCs (2003–04)

	Export composition		Destination (% by category)			
	(10^6 US $)	(%)	EU	MPC	US	ROW
Agricultural products	6,776	100	47	10	5	39
Vegetables and fruit	2,257	33	56	9	5	30
Fish, crustaceans, molluscs	1,020	15	72	3	2	23
Cereals, cereal preprtns	521	8	8	39	3	49
Fixed veg. fats and oils	492	7	73	4	7	16
Crude animal, veg. matl.	480	7	70	2	9	19
Animal, veg. fats, oils, nes	399	6	18	1	1	79
Live animals	285	4	2	14	0	83
Misc. edible products	272	4	29	4	6	60
Sugar, sgr preprtns, honey	172	3	50	15	4	31
Beverages	170	3	26	13	5	55
Dairy products, bird eggs	161	2	4	17	4	75
Coffee, tea, cocoa, spices	144	2	21	19	11	49
Tobacco, tobacco mnfct.	114	2	17	1	16	67
Animal feedstuffs	102	1	11	16	0	73
Meat, meat preprtns	71	1	48	5	7	40
Oil seed, oleaginous fruit	49	1	71	9	3	17
Cork and wood	33	0	61	9	0	30
Hides, skins, fur skins, raw	19	0	13	21	0	66
Animal oils and fats	15	0	21	0	0	79

Note: For Lebanon data from 2003 are used.

Source: International Trade Centre (UNCTAD/WTO) data.

Table 2.5 Value and origin of agricultural imports by MPCs (2003-04)

	Import composition		Origin (% by category)			
	(10⁶ US$)	(%)	EU	MPC	US	ROW
Agricultural products	*16,643*	*100*	*29*	*5*	*15*	*51*
Cereals, cereal preprtns	4,817	29	23	5	30	42
Cork and wood	1,323	8	46	0	1	53
Dairy products, bird eggs	1,287	8	56	3	1	40
Vegetables and fruit	1,251	8	29	15	7	48
Fixed veg. fats and oils	1,223	7	18	2	9	71
Sugar, sug. preprtns, honey	1,003	6	37	2	1	61
Animal feedstuffs	952	6	14	3	19	64
Coffee, tea, cocoa, spices	745	4	20	4	1	75
Tobacco, tobacco mnfct	704	4	16	1	32	51
Oil seed, oleaginous fruit	630	4	4	1	30	65
Meat, meat preprtns	623	4	11	1	4	85
Misc. edible products	571	3	55	10	18	18
Fish, crustaceans, molluscs	422	3	29	7	3	61
Live animals	393	2	52	16	2	30
Crude animal, veg. matl.	295	2	60	5	8	27
Beverages	198	1	71	11	2	15
Animal, veg. fats, oils, nes	166	1	38	3	12	47
Hides, skins, fur skins, raw	31	0	65	9	2	24
Animal oils and fats	10	0	19	0	42	38

Note: For Lebanon data from 2003 are used.

Source: International Trade Centre (UNCTAD/WTO) data.

2.2.3   *Liberalisation achieved by the EMP*

As far as establishing an FTA is concerned, the absence of a defined prospect for the liberalisation of agriculture must be stressed (Barcelona Declaration, European Commission, 1995, emphasis added):

[T]aking as a starting point *traditional trade flows*, and *as far as the various agricultural policies allow* and with due respect to the results achieved within the GATT negotiations, trade in agricultural products will be progressively liberalized through reciprocal preferential access among the parties.

A rather sceptical interpretation of the Barcelona Declaration is that there will be no liberalisation of agriculture, apart from the commitments made within the negotiations on the General Agreement on Tariffs and Trade (GATT), which cannot be withheld from most MPCs by the most-favoured nation principle. This interpretation seems to fit observed behaviour: there have been no significant new concessions by the EU for agricultural products in the EuroMed Agreements, nor are these expected to come about in the near future (Garcia-Alvarez-Coque, 2002, p. 402).

For the MPCs, a preferential treatment of imports originating in the EU is a brand new feature. Compared with EU concessions, MPC preferences are even more limited, in terms of the share of preferential over total trade flows and in terms of tariff reductions. The products concerned are largely staple foodstuffs or 'continental' products.

In contrast to agricultural products, explicit time frames for phasing out protection for manufacturing products are part of the EuroMed Agreements. Whether these will be followed remains an open question, with most MPCs not yet being required to cut back protection. Tunisia, being the first MPC to sign a EuroMed Agreement, is ahead of the other MPCs in reforming its economy. The expectation for Tunis was that an overnight liberalisation of manufactured goods would result in the bankruptcy of one-third of its industrial firms. An adjustment programme has therefore been implemented to prepare firms for increased competition (Riess et al., 2001). In spite of pressure to push back the deadline of 2008 for full liberalisation, Tunisia has been reducing its barriers to European manufactured goods. The expected demise of manufacturing has not materialised and Tunisia has been able to maintain a GDP growth rate of around 5%.

2.2.4 Summarising the liberalisation achieved by the EMP

The above analysis indicates ample scope for liberalisation to have an impact on Mediterranean trade flows. Currently, significant protection exists across all sectors in the MPCs and mainly for agriculture on the EU side. Regardless of the current levels of protection the EU is already a major destination for MPC exports and a major origin for MPC imports. A reduction of trade barriers can thus be expected to have a significant impact on trade flows. The liberalisation achieved through the EMP is, however, limited. Some concessions have been made, but the consensus is that not much can be expected in terms of agricultural concessions from the EU.

A new element of the EMP relative to earlier agreements is the reciprocal character of the preferences, requiring a reduction in the high current protection levels of the MPCs. Sizeable reductions in protection will require a major restructuring of their economies, which is reflected in the availability of MEDA funds to facilitate restructuring.

Before turning in more detail to the patterns of protection on the MPC side, we first review the structural features of the MPC economies. These features affect the impact of trade liberalisation on economic growth and the scope for liberalisation on the MPC side.

2.3 MPC economies

The MPC economies cover a broad spectrum. Where Israel has relatively high per-capita income levels (comparable to EU incomes), the other MPCs included in this study are found in the middle to low income brackets. Correspondingly, diverse models of economic development are found, varying from countries increasingly participating in the world economy to countries with marked protectionist tendencies. In spite of this diversity it is still possible to identify certain common elements across the MPCs, especially those classified as developing countries: high population growth, lagging economic growth, the importance of agriculture and high trade-protection levels.

This section discusses the structure of MPC economies in order to assess the extent to which liberalisation through the EMP may contribute to economic growth. We start with the main characteristics of the MPC economies and their development over time. We then discuss the causes of sluggish economic growth in the MPCs, assessing in general terms the possible impact of the EMP on enhancing growth. We conclude the discussion of the MPC economies by taking a closer look at the agricultural sector and agricultural policies pursued by MPCs.

2.3.1 Key characteristics of MPC economies

Table 2.6 presents key economic indicators of the MPCs. The MPCs are small economies compared with the EU. The total GDP of all the MPCs combined is about the same as the GDP of Spain and some 15% less than the total GDP of the EU accession countries (European Commission, 2003). In general the MPC economies can be characterised by high levels of debt, relatively high population growth and high inflation rates.

Table 2.6 Economic indicators of MPCs (2003)

	GDP level	GDP growth 1990–2000	GDP/ capita	Debt	Population growth
	($ billion)	(%)	(1,000 $)	(% of exports)	(%)
Algeria	61.0	2.3	1.9	n.a.	1.6
Egypt	109.6	4.1	1.6	152.3	1.8
Israel	115.7	4.2	17.3	n.a.	1.8
Jordan	9.6	4.9	1.8	164.5	2.6
Lebanon	17.7	6.0	3.9	603.4	1.3
Morocco	38.5	2.8	1.3	136.0	1.6
Syria	19.7	4.7	1.1	270.7	2.3
Tunisia	21.9	4.6	2.2	155.3	1.2
Palestinian Territories	2.9	n.a.	0.8	n.a.	4.1

Source: World Development Indicators (World Bank, 2005).

High population growth requires strong economic growth to maintain employment. Lack of employment has repercussions for social stability and the scope for structural changes in the economy. Table 2.7 presents demographic data for the MPCs. A comparison of population growth in 2003 in Table 2.6 with annual population growth over the period 1990-2003 in Table 2.7 suggests that population growth is slowing down in most countries. Yet because of previous growth levels, the labour force is continuing to swell rapidly in most MPCs, while unemployment levels are already high (up to a third of the labour force in Algeria).

From these aggregate data a picture emerges of economies that have not been able to expand quickly enough to absorb a fast-growing labour force. At the same time, most of the countries face serious debt problems, reducing the room for economic restructuring. The positive exception in economic terms is Israel, faring much better than the surrounding countries. The next sub-section considers the reasons for the lagging economic growth in the majority of the MPCs.

Table 2.7 Demographic characteristics of the MPCs (2004)

	Population		Labour force		Rural population	Unemploy-ment
	Total	Growth[a]	Total	Growth[a]		
	(mn)	(%)	(mn)	(%)	(%)	(%)
Algeria	31.8	1.9	11.7	4.0	41.2	29.8
Egypt	67.6	2.0	26.7	3.0	57.2	11.2
Israel	6.7	2.8	2.9	3.6	7.9	9.1
Jordan	4.5	4.0	1.7	6.2	20.9	25.0
Lebanon	30.1	1.7	1.7	3.1	9.4	20.0
Morocco	17.4	1.7	12.2	2.4	42.6	18.2
Syria	9.9	2.8	5.8	4.2	47.5	20.0
Tunisia	3.4	1.5	4.2	2.9	32.6	14.3
Palestinian Territories	5.3	4.2	0.7	5.1	n.a.	>30

[a] Annual growth 1990-2003.

Sources: World Bank (2005); unemployment data from the European Commission (retrieved from http://ec.europa.eu/comm/external_relations/euromed).

2.3.2   Causes of lagging economic growth

High population growth and extensive unemployment make economic growth a prime issue for the MPCs. But the actual track record of the MPCs is rather disappointing, lagging behind the growth rates attained by comparable countries in other parts of the world. Figure 2.2 sketches different, interconnected causes of the sluggish economic growth in the MPCs, generalising over the different countries. The three main forces hampering growth are non-trade income, high levels of trade protection and extensive state interference in the economy.

A first factor slowing economic growth is the presence of non-trade income. Oil exports and remittances are important sources of foreign exchange for a number of countries. This inflow of foreign exchange distorts the economy (the so-called 'Dutch disease') by boosting domestic demand. The resulting appreciation of the exchange rate promotes investments in non-traded sectors of the economy, while reducing investments in the traded sectors. Such a distorting role of oil income is relevant in Egypt, Algeria and Syria (Riess et al., 2001), while the ratio of remittances to the value of exports approaches this scenario in Egypt and Jordan, underscoring the element of (temporary) migration in the MPC economies (Nassar & Ghoneim, 2002).

Figure 2.2 Outline of the causes of slow economic growth in the MPCs

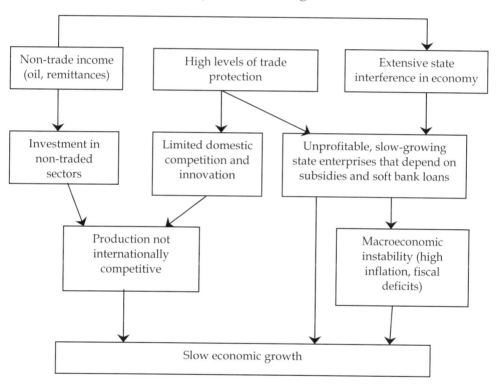

A second factor undermining economic growth is the high level of trade protection through an inward-looking development strategy. Such protection shelters domestic firms from international competition. This in turn reduces incentives for efficiency improvements and investments in innovations. Combined with the focus on non-traded sectors induced by inflows of non-trade foreign exchange, this approach has resulted in a production structure that is not internationally competitive.

A third factor hampering economic growth is the extended influence of the state on the economy. Such influence takes the shape of an over-staffed public sector and a dominant presence of state enterprises. The bloated character of public sector employment is apparent from the fact that its share in (non-military) employment is twice the global average, accounting for close to one-fifth of employment in the MPCs. On the production side the state also plays a significant role, for example accounting for 30% of GDP in Egypt and Tunisia, and close to 60% in Algeria. Public investments were close to 40% of total investment, which is double the middle-income country average. Booming oil revenues in the 1970s and 1980s provided a further stimulus to increasing public sector expenditures in oil-exporting countries (Bulmer, 2000; Riess et al., 2001).

High levels protection and extensive government involvement in the economy has led to unprofitable state enterprises maintained with subsidies and soft loans. This policy puts heavy pressure on state budgets, causing severe macroeconomic instability. Thus reforms are about as necessary as they are difficult. Tariff revenues form an important part of the government budgets. Countries with high import ratios from the EU will be faced with significant losses in income if they grant the EU preferential access to their markets, necessitating a restructuring of their economy.

2.3.3 The structure of the agricultural sector and MPC agricultural policies

Although agriculture is an important sector across the board, its role in the overall economy differs by MPC (see Table 2.8). The classification of most MPCs as developing countries is reflected by the large share of the labour force employed in agriculture (between 12 and 36% of the total labour force). In high-income countries this proportion is less than 10%, as reflected by the 2% figure for Israel.

In terms of GDP, agriculture is a major sector, again mostly for developing MPCs, contributing as much as 23% to GDP in Syria. Jordan seems to have a more particular economic structure, although agriculture contributes significantly to employment (accounting for 12% of employment); its contribution in terms of the GDP share is limited to 3%.

Table 2.8 Contribution of agriculture to GDP and employment in the MPCs

	Share of GDP (%)			Agricultural labour force (% total labour force)	
	Agriculture	Industry	Services	1990	1999
Algeria	21	13	66	26	24
Egypt	19	23	58	40	33
Israel	2	32	66	4	3
Jordan	3	19	79	15	12
Lebanon	14	12	74	7	4
Morocco	22	19	59	45	36
Syria	23	26	51	32	28
Tunisia	13	20	67	29	25
Palestinian Territories	13	14	73	n.a.	n.a.

Notes: GDP data are for 2003 (2002 in the case of Egypt); an exception is the GDP by sector for Israel (2003 estimates from the CIA World Factbook).

Sources: World Development Indicators, World Bank (2005), FAOSTAT and the CIA World Factbook (retrieved from https://www.cia.gov/cia/publications/factbook/geos/is.html).

Production composition varies considerably from country to country, but it centres on cereals, fruit and vegetables, followed by other staple foodstuffs or typically Mediterranean products. The importance of agricultural raw materials, such as tobacco, cotton and sugar beet is more limited. Egypt is a main producer, especially for staple foodstuffs, although other countries – Israel, Syria and those in the Maghreb[6] – are often important players in particular markets.

In addition to the fundamental climatic and geographical features of the Mediterranean area, the composition of production is affected by long-term trends in world prices and relatively lower levels of protection for some target markets for Mediterranean products (fruit and vegetables, olive oil). Products that are not strictly Mediterranean (cereals, meat and

---

[6] The Maghreb region covers Morocco, Algeria and Tunisia.

milk) maintain an important role in the agricultural system, by providing subsistence to peasant farmers and in some cases, because of policy support and trade protection aimed at reducing dependence on imported food (INEA, 2002; De Rosa, 1997).

The main issues in MPC agriculture can be summarised under three headings: a polarised production structure, production limitations and food security. There is a marked and growing polarisation between large-scale, capitalist company farming and small family holdings. Institutional factors, insufficient public intervention in the reform of land ownership and, for some MPCs, strong government support for agricultural exports have accentuated this duality. Large farms complain above all about the lack of adequate services, while small farmers find it difficult to make a living from traditional farming practices. These issues stem from natural and technical restraints, obstacles to mechanisation and other structural limitations, but also from price dynamics and conditions of the marketing channels (INEA, 2002).

Environmental, climatic and technological limitations restrict the expansion of arable land and create problems for the sustainability of traditional agricultural methods and ecosystems. The lack of fertile land and water is an evident hindrance to agricultural development, while the goal of increasing yields creates further problems, owing to chemical inputs already being used on a massive scale. Desertification, soil erosion and infertile soils are serious problems brought about by overgrazing, intensive crop rotation and the abandonment of traditional agricultural practices. Inefficient and insufficient consideration of soil characteristics are often a feature of the management of water resources and can lead to the soil becoming too saline or alkaline, as happens in Syria and Egypt, or to soil erosion, as is widespread in Syria, Lebanon and the Maghreb (Makhlouf et al. 1998; Lacirignola & Hamdy, 1995).

The orientations and tools of agricultural policy differ from country to country.[7] In the case of Israel, agricultural policy has been influenced by a need to combine agricultural development with national security and self-sufficiency in food production, given its hostile geopolitical environment. As regards the other MPCs, the major priorities in agricultural policy are to improve the performance of the sector and the

---

[7] The remainder of this section is based on INEA (2002).

level of food security. Minor objectives – but by no means negligible – relate to improving linkages between vertical stages of agro-food systems (competitiveness, marketing, etc.), as well as environmental protection, food quality and food safety.

Measures of producer support and market regulation evolve slowly within the context defined by adjustment programmes, WTO commitments and preferential deals with the EU. All imply, for developing Mediterranean countries, a fundamental change in price policies, with the aim of restoring the market mechanism and improving its operation. Liberalisation and structural adjustment have important implications for agriculture. Agricultural policy reform seeks to open domestic markets, reducing the protection differentials among agriculture and other sectors, along with scaling back government support for production prices or cutting consumer subsidies (or both) and slashing input subsidies.

Some countries have made considerable strides in this direction. Egypt as well as the Maghreb countries modified their policies appreciably in order to reduce protection in the industrial sector and re-launch agriculture by improving market efficiency. The effects of the reforms vary depending on the starting point of the country concerned and the level of social consensus, but the overall picture is still characterised by hefty government regulation of agricultural markets, through intervention on prices (consumer and producer subsidies), quantities (quotas) and tariffs. In fact, the reform process has been rather selective: government support and trade protection are still considered indispensable for certain products, while policy interventions aspiring to either control food prices or extract surpluses from the agricultural sector are still in place. Moreover, there are still cases in which agriculture suffers from an overvaluation of real exchange rates and trade protection in the manufacturing sector.

2.3.4   EuroMed Agreements and migration

Since remittances are a key feature of the MPCs, one can wonder about the impact of the EuroMed Agreements on migration from the MPCs to the EU. This question is largely absent from existing studies of the EuroMed Agreements, although reducing the flow of North African migrants is an important policy issue for the EU. The EuroMed Agreements could reduce migration if they were to stimulate labour-intensive production in the MPCs, thus increasing wages and reducing the incentives for migration. This standard argument assumes that trade and migration are substitutes.

Preliminary analyses of the impact of the North American Free Trade Agreement on migration flows between Mexico and the US, however, suggest that the establishment of an FTA can increase migration flows. Different theoretical explanations can be provided for such an outcome. One is that a rise in wages following the FTA has enabled low-skilled labourers to accumulate the cash needed for migration. In this line of thought, trade and migration are complements, not substitutes. New trade theory, allowing for a concentration of economic activity, provides an alternative explanation. Reducing trade barriers opens the way for industries to become more concentrated. Given the hub-and-spoke structure of the EuroMed Agreements this concentration is likely to occur in the EU. The establishment of an FTA could thus increase migration.

2.3.5 Summarising the main features of MPC economies

The still rapidly-growing labour force in the MPCs necessitates the acceleration of economic growth. The track record of the MPCs is not promising and growth rates have been lagging compared with other regions, owing to inflows of foreign exchange from oil and remittances, high trade-protection levels and prolific state interference. State intervention in the agricultural sector has also been substantial, through price interventions, quotas and tariffs. The next section takes a closer look at the different forms of protection in the MPCs.

2.4 Scope for liberalisation by the MPCs

A key element of the EMP is the reciprocal nature of the trade liberalisation. The scope for liberalisation by the MPCs depends on the current level of protection and political possibilities for liberalisation. Most studies of the EMP focus on the (lack of) liberalisation on the EU side. In terms of the impact on the MPCs, the lagging liberalisation of agricultural trade by the EU is most important since barriers to manufactured exports to the EU were lifted in the 1970s. The position of the EU with respect to the liberalisation of agricultural trade is well-documented and there is no reason to expect a major change to this position. Thus trade protection by the EU will not be further discussed here.

This section examines the trade barriers erected by the MPCs. Although protection is only one of a multitude of hindrances to economic growth in the MPCs, it is the aspect most easily influenced through the EMP. Furthermore, an essential feature of the trade liberalisation in the context of the EMP is its reciprocal nature, requiring the MPCs to remove their trade barriers as well. This feature of the EMP also warrants a closer look at the protection in the MPCs.

First, trade protection by the MPCs is compared with other regions and its development over time is assessed. Second, non-trade polices affecting production and trade patterns are discussed. The third sub-section addresses some of the reasons for the high levels of trade protection and other interventions in the MPCs.

2.4.1 *Comparing trade protection in the MPCs with other regions in the world*

Detailed studies of protection by the MPCs are few and far between. Srinivasan (2002) provides a rare summary of different studies, of which the main points are summarised here. Table 2.9 reproduces protection indicators for a selected number of MPCs, for which data are available in the Trade Analysis and Information System (TRAINS) database. Table 2.10 reproduces the scores of the MPCs on three aggregated protection measures.

All indicators confirm the observation previously made about high trade-protection levels in the MPCs. The weighted average tariff rate provides an illustration of the protection rates. Egypt has the lowest score of the MPCs with a weighted tariff of 13.7%. This rate still exceeds the highest score by income group (12.6% for low-income countries). This result reflects the lack of trade liberalisation in Egypt while the rest of the world has been lowering trade barriers. The MPCs started from historically high protection levels and have not been reducing (sometimes even increasing) them.

Among the MPCs, Egypt, Morocco and Tunisia have gone farthest in reforming their economies. This reform, however, has not extended to trade protection. This situation is illustrated by the weighted tariff of Tunisia being more than double the tariff of the low-income countries.

In terms of aggregate protection measures the MPCs score even worse. The aggregate measures build on the simple average tariff rate and the standard deviation, both of which are exceptionally high in the MPCs. In terms of the Andrew and Neary measure of trade protection, only South Asia comes near the protection levels of the MPCs. Within the group of MPCs included in Tables 2.9 and 2.10, Jordan has the lowest level of protection, owing to extensive reforms in the 1990s.

Table 2.9 Trade protection indicators for selected Middle East and North African countries

	Simple avg. (%)	Weighted avg. (%)	Standard deviation (%)	NTB coverage (%)	Escalation index (Ratio)
Selected MPCs					
Algeria	24.2	17.3	16.7	15.8	1.6
Egypt	28.1	13.7	130.6	28.8	2.1
Jordan	21.6	18.9	15.8	0.0	1.1
Morocco	35.7	25.8	31.2	5.5	1.1
Tunisia	29.9	28.8	12.8	32.8	1.1
Comparators by income group					
Low-income countries	15.5	12.6	10.9	5.5	1.5
Lower middle-income countries	15.3	12.5	15.0	13.4	1.7
Upper middle-income countries	13.8	11.6	12.3	14.7	1.6
High-income countries	4.3	3.4	7.0	15.6	1.7
Comparators by region					
Europe and Central Asia	9.8	6.7	11.0	10.9	2.0
East Asia	13.1	8.7	16.8	9.9	1.8
Latin America	13.1	11.9	8.5	17.1	1.6
Sub-Saharan Africa	17.7	14.2	13.3	4.5	1.5
South Asia	19.7	18.8	11.7	8.2	1.2

Note: Most-favoured nation tariff rates are used; NTB coverage refers to the percentage of tariff lines having at least one non-tariff barrier; the escalation index is the ratio of the simple average tariff on final goods to the tariff on intermediate goods.

Source: Table 1 in Srinivasan (2002, p. 9) based on the TRAINS database.

Table 2.10 Aggregated trade protection indicators for selected Middle East and North African countries

	Sharer index	Oliva index	Anderson & Neary index
Selected MPCs			
Algeria	7.0	20.0	25.0
Egypt	8.0	55.8	23.5
Jordan	4.0	14.4	n.a.
Morocco	7.0	26.6	35.0
Tunisia	8.0	26.1	23.8
Comparators by income group			
Low-income countries	3.5	11.7	21.2
Lower middle-income countries	4.1	14.7	15.1
Upper middle-income countries	4.1	13.6	11.8
High-income countries	3.1	8.0	10.9
Comparators by region			
Europe and Central Asia	3.5	10.4	11.6
East Asia	3.9	13.2	11.3
Latin America	3.6	12.9	14.7
Sub-Saharan Africa	3.8	13.1	18.9
South Asia	4.2	14.6	27.7

Note: The Sharer index is an arbitrary scoring system combining most-favoured nation tariff rates and standard deviations; the Oliva index combines tariff rates, standard deviations and non-tariff barrier coverage; the Anderson and Neary index is the uniform tariff rate applied to a free trade regime to return welfare to the most recent year of observation.

Source: Table 1 in Srinivasan (2002, p. 9) based on the TRAINS database.

The EMP could be expected to lead towards a more open orientation of the MPC economies. Table 2.11 shows the development over time of the trade restrictiveness indicator. In 2003 all the MPCs, except Israel, can be classified as having restrictive trade policies. More interesting is the development over time. One could have expected more outward-looking trade policies as time went by. According to Table 2.11, only Egypt and Syria seem to fit this pattern, moving to a more open trade policy in 2003. The patterns of the other countries, however, suggest that this may only be a temporary change. Jordan, for example, increased its protection in 2003. Morocco and especially Lebanon alternate between open and restrictive trade policies. Overall, there is no indication that the MPCs are following the global trend towards more open policies.

Table 2.11 Trade restrictiveness over time by country

	1995	1996	1997	1998	1999	2000	2001	2002	2003
Algeria	5	5	5	5	5	5	4	4	5
Egypt	5	5	5	5	5	5	5	5	4
Israel	2	2	2	2	2	2	2	2	2
Jordan	4	4	4	4	4	4	4	4	5
Lebanon	n.a.	2	2	5	2	5	3	4	5
Morocco	5	4	4	5	4	4	4	5	5
Syria	n.a.	5	5	5	5	5	5	5	4
Tunisia	5	5	5	5	5	5	5	5	5

Note: Restrictiveness index: less than 2 = free trade policy, between 4 and 5 is repressed trade policy.

Source: Heritage Foundation (Gwartney et al., 2003).

2.4.2   Government intervention in agriculture

Apart from restrictive trade policies consisting of tariffs, licensing, import bans, state trade monopolies, multiple exchange rates and restrictive foreign-exchange allocations, other MPC policies may affect domestic production and the scope for foreign competitors (ERF, 2002).

Agriculture is important for the MPCs, in terms of employment, contribution to GDP and income-distribution effects (poverty tends to be concentrated in the rural regions). Historically, the MPC governments intervened in the agricultural sector through purchases of commodities (cereals, vegetable oil and sugar), output support for both farmers and agro-food processors, subsidies of inputs (water, fertilisers, seed and machinery) and consumer subsidies. The amount of intervention varied with the commodities, from directly controlling prices and markets for strategic commodities (such as wheat and sugar beet) to allowing competitive markets for fruit and vegetables. The policy interventions require huge outlays of public funds and are hard to maintain when economic circumstances take a change for the worse. Some countries reformed their agricultural interventions in the course of implementing structural adjustment programmes (Algeria, Jordan, Morocco and Tunisia); others (notably Syria) have maintained the prime role of the state (Chaherli, 2002).

Despite reforms, distorting policies remain. On the production side these mainly affect 'strategic products': wheat in most MPCs; milk and olive oil in Tunisia; cotton, sugar beet and tobacco in Syria; and sugar beet and tobacco in Lebanon (Chaherli, 2002). On the input side a wide array of subsidies are used to stimulate cereal and livestock production. Subsidies of feed area is a common feature, representing the most important (and constant) item of the agricultural budget in MPCs. Analyses of the competitiveness of MPC production indicate that in the absence of government intervention in output and input prices, MPC producers could not compete with foreign imports.

High tariffs to ward off cereal imports and to promote domestic production lead to high domestic prices. Most of the MPC governments subsidised consumer prices to reduce the burden for consumers, but found that intervening on both sides of the market poses a heavy burden on government budgets. Therefore, most countries have since reduced food subsidies, generally in a gradual fashion to avoid political instability.

Apart from tariffs and quotas there seem to be other restrictions to trade since quota fill rates are (well) below 100%. It seems likely that the administration of the quotas is hampering foreign imports. In the case of Egypt, for example, product standards are applied to imports that do not apply to domestic products, such as the requirement that imported beef has less than 7% fat content while no restrictions apply to domestic beef (USTR, 2003).

2.4.3 Factors affecting the slow pace of reform in the MPCs

The above comparison of MPC protection with that in other regions begs the question of why the MPCs are not following the global trend towards freeing trade. The implication of diverging from the global pattern is that the MPCs will be losing in terms of international competitiveness relative to other regions in the world. Thus there seems to be a clear case for at least keeping protection levels in line with the rest of the world.

Analysis of the relative strength of import-substitution industries and export-oriented industries in Morocco and Egypt suggests a strong lobby for maintaining protection rates (Srinivasan, 2002). This leaves the question of why there is such a strong import-substitution industry. Trade policies unfavourable to exports and generally unsupportive domestic policies seem to play a role in the development of import-substitution industries demanding continuous protection.

Policies oriented towards import substitution were made possible by the inflows of foreign exchange (from natural resources and remittances) and by the preferential agreements in which the MPCs engaged. Income from fuel and remittances allowed the financing of an expansive public sector and of the interventions on the producer and consumer sides. At the same time a number of MPCs enjoyed preferential access to EU markets for textiles and clothing. The resulting investments in these sectors may have been at the expense of investments in medium technology industries (such as industrial chemicals, standardised machinery and simple electrical and electronic products). Latin American countries (with similar endowments) did make these investments, gaining market share. Given the learning-by-doing that plays an important role in manufacturing, the inward-looking strategy of the MPCs in the past could prove costly in the future.

2.4.4 Summarising the scope for liberalisation by the MPCs

Historically MPCs have had high levels of protection and diverged from the global trend towards more open policies. In addition to restrictive trade policies there is still a hefty degree of government intervention in the MPC economies. Two major factors that hamper MPC trade liberalisation are a bloated public sector and the lack of international competitiveness of MPC producers. Liberalising trade would reduce government income and may wipe out a large part of MPC production, both of which may result in massive social unrest. Given past trends and the current economic context in the MPCs, the reciprocal trade liberalisation implied by establishing a free trade agreement seems hard to achieve.

The impact of preferential access for textiles and clothing on industrial developments raises a question about the impact of preferential access granted by the EU for specific agricultural products. Similar to the distortion in manufacturing, these preferences may have distorted the agricultural production structure of the MPCs.

Conclusions

The current implementation of the EMP can be summarised as a trade-focused agreement in which no significant results have been attained in the reduction of agricultural trade protection, either by the EU or by the MPCs. Although a reduction of protection on manufactured goods by the MPCs is specified by time schedules, previous trends give no reason to expect that these schedules will be met. Rising trade with the US suggests that American trade agreements and financial support are more effective than the present European initiative.

The lack of progress in liberalising agricultural trade indicates that the economic interests of a limited set of (Mediterranean) EU member states has so far prevailed over the overall interests of the EU in economic growth at its southern borders. Thus, EU foreign and trade policies towards the MPCs are currently incoherent. The explicit consideration of these conflicting objectives could support the development of alternative policies more in line with the multiple objectives of the EU. In this light the recently drafted road map for liberalising agricultural trade based on a negative-list approach (IPTS, 2006) appears to be a promising sign for a realignment of the EU's trade policies with the EU's overall interests.

The liberalisation of trade addresses only one of the features of MPC economies that hamper their growth. Liberalising trade not only reduces high trade-protection levels in the MPCs, but also has strong implications for employment and government tariff revenues. Taking account of the structural aspects of the MPC economies could facilitate the design of agreements such that they promote growth and support a gradual move towards more open economies, while acknowledging the constraints faced by the MPC governments.

Finally, the EMP is designed as a set of bilateral agreements between the EU and each of the MPCs. The Mediterranean FTA sought by the EMP, with the view that linking the MPC economies could play an important stabilising role in the region, requires the liberalisation of trade among the MPCs as well. The current set of agreements, however, results in a hub-and-spoke structure – which does not promote South–South integration.

References

Bulmer, E.R. (2000), *Rationalizing Public Sector Employment in the MENA Region*, Working Paper No. 19, World Bank, Washington, D.C.

Chaherli, N. (2002), *Agricultural Trade Liberalization: Main Issues for the MENA Region*, Middle East and North Africa Working Papers, World Bank, Washington D.C.

Central Intelligence Agency (CIA) (2002), *CIA World Factbook 2002*, CIA Office of Public Affairs, Washington, D.C.

Dell'Aquila, C. and M. Kuiper (2003), *Which Road to Liberalization? A First Assessment of the EuroMed Association Agreements*, ENARPRI Working Paper No. 2, CEPS, Brussels, October.

De Rosa, D.A. (1997), *Agricultural Trade and Rural Development in the Middle East and North Africa: Recent Developments and Prospects*, World Bank Policy Research Working Paper No. 1732, World Bank, Washington, D.C.

Economic Research Forum (ERF) (2002), *Economic Trends in the MENA Region 2002*, ERF, Cairo.

European Commission (1995), The Euro-Mediterranean Partnership – Barcelona Declaration adopted at the Euro-Mediterranean Conference, 27-28 November 1995, Brussels.

———— (1997), *Study on the impact of Mediterranean concessions*, COM(1997), 477 final, European Commission, Brussels.

———— (2001), *Annual report of the MEDA programme 2000*, COM(2001), European Commission, Brussels (retrieved from www.europa.eu.int/comm/external_relations/euromed/meda/reports.htm).

———— (2003), *Economic Review of EU Mediterranean Partners*, European Economy Occasional Paper No. 2, European Commission, Directorate-General for Economic and Financial Affairs, Brussels.

European Council (2003), *A Secure Europe in a Better World*, European Security Strategy presented by Javier Solana, Brussels, 12 December.

Garcia-Alvarez-Coque, J.-M. (2002), "Agricultural trade and the Barcelona Process: Is full liberalization possible?", *European Review of Agricultural Economics*, Vol. 29, No. 3, pp. 399-422.

Gwartney, J., R. Lawson and N. Emerick (2003), *Economic Freedom of the World: 2003 Annual Report*, Fraser Institute, Vancouver, B.C. (retrieved from www.freetheworld.com).

Istituto Nazionale di Economia Agraria (INEA) (2002), *L'Unione Europea e i paesi terzi del Mediterraneo: Accordi commerciali e scambi agroalimentari*, Osservatorio sulle Politiche Agricole dell'UE, INEA, Rome.

Institute for Prospective Technological Studies (IPTS) (2006), *Euro-Med Association Agreements: Agricultural Trade – Regional Impacts in the EU (Proceedings of the Workshop)*, European Commission, Directorate-General Joint Research Centre, Brussels and IPTS, Seville.

Lacirignola, C. and A. Hamdy (1995), "Introduction" in "Focus: Land and water management in Mediterranean countries", *MEDIT*, No. 1/95.

Makhlouf, A., M. Abdelouahab and M. Aziz (1998), "L'eau dans la region del L'Afrique Nord et du Moyen Orient: de la penurie al la securité", *MEDIT*, No. 2/98.

Nassar, H. and A. Ghoneim (2002), *Trade and Migration, Are they Complements or Substitutes: A Review of Four MENA Countries*, paper presented at the "Annual Bank Conference on Development Economics", 24-26 June, Oslo, Center for Economic & Financial Research and Studies, Cairo.

Riess, A., P. Vanhoudt and K. Uppenberg (2001), "Further Integration with the EU: Just One Ingredient in the Reform Process", *EIB Papers*, Vol. 6, No. 2, pp. 58-86.

Srinivasan, T.G. (2002), *Globalization in MENA – A Long Term Perspective*, paper presented at the "Fourth Mediterranean Development Forum", held in Amman, Jordan on 6-9 October, World Bank, Washington D.C.

United States Trade Representative (USTR) (2003), *National Trade Estimate Report on Foreign Trade Barriers*, Office of the United States Trade Representative, Washington D.C.

World Bank (2005), World Development Indicators, World Bank, Washington, D.C.

3. Reforming the EU's domestic support in the Doha round
Measurement, feasibility and consequences

Hans Jensen and Wusheng Yu

Introduction

The agricultural trade negotiations of the World Trade Organisation (WTO) were put back on track with the agreement to the July package in 2004. The domestic subsidies allocated by a few member countries to their agricultural sectors formed part of the negotiations. Many developing countries had been pressuring these members to reduce their domestic support measures. In responding to these pressures, the July package outlines a framework for such reductions. In this framework, the guidelines for reforming domestic support measures through a tiered formula have been set out but the numerical targets of the formula have not been specified and are still to be negotiated.

Exploring the possible outcomes of the negotiations under this framework and evaluating their impact is no easy task. Among other things, it requires detailed knowledge of the WTO measurements of domestic support programmes and of the actual policy instruments used by individual member countries. These and other specific challenges that analysts face are discussed below.

Unlike the cases of reducing market access barriers and export subsidies, reforming domestic support measures involves a complex package of policy instruments that are placed in different WTO 'boxes' (see Figure 3.1, reproduced from Baffes & de Gorter, 2005).

Figure 3.1 Domestic support measures and the WTO

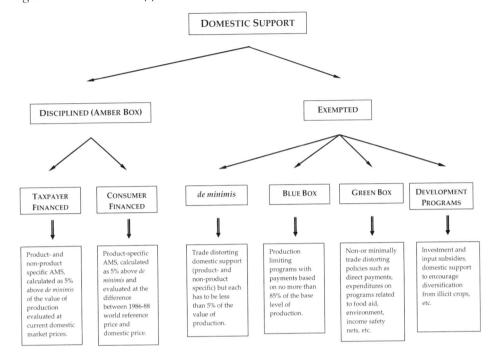

Notes: The 5% above *de minimis* applies for the sum of taxpayer- and consumer-financed support. The *de minimis* exemption can include consumer-financed support.

Source: Baffes & de Gorter (2005).

Any reductions will be applied to the measures in these boxes and to the overall levels of trade-distorting domestic support derived from the different boxes. To complicate matters, the current levels of domestic support for each member country allowed under the Uruguay Round Agreement on Agriculture (URAA) are tied to historical references, such as average production values, land areas, intervention prices and average world market prices. Some of these measures have since lost their relevance because of changing market conditions and (changing) trade policies. For example, the market price support included in the amber box for the EU is largely no longer relevant. Yet the EU is still obliged to include some 'fictional' numbers in its notification of the aggregate measure of support (AMS) to the WTO. What this implies for the talks about reductions is that it becomes quite difficult to figure out the actual level of support using the

WTO classification scheme (i.e. the boxes) and it is even more difficult to come up with a numerical reduction target, given the current levels of permitted support and the actual levels of support. These issues lead to the research question of how to correctly measure domestic support programmes in policy analysis.

Furthermore, a numerical analysis of the effects of any reduction proposal needs to match the support given to individual instruments in each of the boxes with the actual policy variable in an economic model. Researchers typically encounter two possible issues. First, one may have knowledge of the aggregate level of support in each box but not the details associated with individual policy instruments within the boxes. Second, different countries usually have very different policy instruments within the same box. Therefore, a sensible model-based evaluation of a reduction proposal needs to translate the reduction proposal into changes to the actual policy instruments in the model and the implementation of the reduction proposal needs to be conducted on a country-by-country basis.

Lastly, unlike the reductions of tariffs and quotas, reforms of domestic support tend to imply cuts to fiscal expenditures on these programmes rather than reductions of the price wedges. In fact, if the fiscal spending needs to be set at a certain level, the relevant price wedge then needs to be adjusted endogenously in response to a reduction of the support in dollar terms. In many cases, reforms to domestic support programmes may also involve turning a trade-distorting measure into a 'decoupled' instrument. There, analysts need to make sure that the new instrument is indeed decoupled from production decisions about the affected products.

The EU maintains the largest domestic-support programme in the world, which is likely to be subjected to new WTO disciplines and reduction commitments at the conclusion of the Doha negotiations. The above-discussed measurement and modelling challenges are perhaps most evident in analysing the possibility, feasibility and consequences of reforming domestic support in the EU. The EU's efforts to meet its potential WTO domestic-support commitments are in essence the reform of the common agricultural policy (CAP). Recent and ongoing efforts to reform the CAP include Agenda 2000, the mid-term review, reforms to the EU's sugar regime and proposed sectoral initiatives in areas such wine, fruit and vegetables. In addition, the enlargement of the EU also has important implications on domestic support negotiations. Therefore, speculation about the possible stance of the EU in the WTO talks and evaluations of the impact of the EU's position need to be conducted in the context of these CAP reform programmes, taking into account external pressures facing the EU.

The purpose of this chapter is to sketch out a reduction proposal for reducing the EU's domestic support measures in the spirit of the July package framework. The possibility and feasibility of the proposal is analysed by carefully reviewing the allowed level of support for the EU (section 3.1) and the existing and ongoing reform initiatives of the CAP (section 3.2). Based on this, we also project a tiered reduction formula for other WTO members (section 3.3). We then proceed to discuss the modelling of existing EU domestic-support programmes and their reforms as per the reduction proposal in a computable, general equilibrium (CGE) model. There, we show how we have mapped out the instruments in the model according to the WTO boxes and how we have modelled the CAP reform, which in the context of meeting the WTO's reduction proposal largely implies a decoupling of the amber and blue box programmes (section 3.4). After that, the numerically simulated results from implementing the proposal are presented and compared with those obtained by reforms to market access and export competition (section 3.5). This section is followed by a discussion and concluding remarks.

3.1 Domestic support in the WTO negotiations and implications for the EU

In the 1994 URAA, disciplines were introduced to limit agricultural support programmes that encourage agricultural production. Member countries made commitments to bind their levels of domestic support using an AMS framework, which provides a measurement for domestic support classified as amber box policies. Domestic support measures that are not classified in the amber box were exempted from reduction commitments. These measures include trade-distorting domestic support policies falling under the *de minimis* level of support (which is set at less than 5% of the value of production in developed countries), support measures contained in the blue box (where payments are linked to historical production levels) and green box measures that are considered to have minimal trade-distorting effects. An overview of these domestic support measures is shown in Figure 3.1.

In the current round of agricultural trade negotiations, disciplined domestic support has been extended to cover blue box policies and a new overall limit on all trade-distorting domestic support has been proposed. More precisely, the Doha Work Programme (WTO, 2004) – also known as the July package – stipulates a framework to reduce the possible use of amber (AMS) and blue box payments, plus payments falling under the *de minimis* rule. Overall, the framework calls for substantial reductions in trade-distorting domestic support and specifies that special and differential treatment will be an integrated part of domestic support. Furthermore, there is to be a strong element of harmonisation in the reductions made by developed countries.

To secure substantial reductions, the framework proposes to cap both the *de minimis* support and the allowed amount of blue box support. More specifically:

- The overall trade-distorting domestic support (as measured by the final bound total AMS plus the permitted *de minimis* level plus the highest level of blue box payments during a recent representative period) will be reduced according to a tiered formula. In the first year of implementation countries have to reduce support by 20% relative to this overall base.

- The final bound AMS will be reduced substantially, using a tiered approach, with those members having higher total AMS making greater reductions.

- Reductions in the *de minimis* support are also to be negotiated.

- Blue box support in the future will not exceed 5% of a WTO member's average total value of production during a historical period to be agreed.

- Direct payments that do not require production under certain conditions can be placed in the blue box.

The EU is the world's largest provider of domestic agricultural support in dollar terms. The total possible trade-distorting domestic support of the EU specified as the final bound AMS commitment in the URAA exceeds €67 bn, with no limit on blue box payments. Of course, not all this allowed amount has actually been spent by the EU in the years following the Uruguay round. In fact, the notified total AMS of the EU in the year 2001 was €43.6 bn, with an additional €22.2 bn in the blue box.

The framework for reducing domestic support proposed in the July package calls for the EU to make larger cuts to its trade-distorting domestic support measures. In fact, many speculate that the EU will be put into the top tier in a tiered reduction formula. For example, Canada has asked the EU to make the largest proportional cuts in its domestic support ceiling as part of a WTO Doha round agreement and proposed to position the EU alone in the first tier of reduction commitments (Agra Europe, 2005). Canada suggests that the US and Japan should be placed in the second tier, with other developed countries in the third tier and developing countries in the fourth tier.

Facing the pressures from other WTO members to make the largest cuts in its domestic agricultural support, the natural question to ask is whether the EU is willing and in the position to agree to a first-tier cut. In this chapter we argue that the CAP reforms in the past decade have given the EU a wide margin for accommodating a large cut to its domestic support measures. If indeed the EU agrees to a first-tier cut, the pressure would be on other countries with substantial domestic-support programmes to follow suit.

3.2 The CAP reform and possible EU stances in domestic support negotiations

Following the guidelines laid out by the Doha Work Programme, Table 3.1 reveals how large a reduction in total trade-distorting domestic support the EU-15 can take from its overall base-level commitment, taking into account the recent Agenda 2000, the mid-term review (MTR) and the sugar and olive oil sector reforms to the CAP.

In the EU-15 the overall base level of all trade-distorting domestic support is assumed to be €112,874 mn, which has been calculated by adding the URAA final bound AMS levels together with the permitted *de minimis* payments in a given reference period plus the highest of existing blue box payments during the 1995-2002 period.[1] Comparing the overall base level of trade-distorting domestic support with the notification to the WTO for the year 2000–01, the EU-15 was already well below this base

---

[1] The permitted *de minimis* level is simply calculated as 5% of the average value of total agricultural production during the period 1999-2002 using values found in the OECD's tables on producer support estimates (PSEs).

commitment of €112,874 mn for that year: the reported AMS plus *de minimis* and blue box payments only amounted to €66,455 mn. Moreover, the recent CAP reforms will further reduce the actual amount of trade-distorting support in the future, implying that large cuts from not only the total base level but also from the actual notification level are possible. Next, we go through the possibilities of reducing the EU support measures contained in the AMS, the *de minimis* support and the blue box.

Table 3.1 The EU's base commitments in trade-distorting domestic support and possible reductions (mn €)

| | AMS | De minimis | | Blue box | Total trade-distorting domestic support |
		Non-product specific	Product-specific		
Base commitment	67,159	12,097	12,097	21,521	112,874
Notif. 2000–01	43,654	538	40	22,223	66,455
Of which MPS [a]	30,684	0	0	0	n.a.
Reductions					
MPS reductions to					
Rice	-376	0	0	0	0
Rye	-238	0	0	0	0
Other cereals	-1,701	0	0	0	0
Milk	-1,893	0	0	0	0
Beef	-11,190	0	0	0	0
Sugar	-6,090	0	0	0	0
Olive oil	-2,070	0	0	0	0
Total MPS reductions	-23,558	0	0	0	-23,558
Non-MPS reductions	17	0	-17	-18,223	-18,223
New domestic support [b]	20,113	538	23	4,000	24,674
New commitment [b]	20,148	6,049	6,049	12,097	28,219

[a] MPS refers to market price support.

[b] This is a conservative estimate of the EU's possible reduction in notified trade-distorting support given that reforms of the cotton, tobacco and hops sectors have not been taken into account in this calculation. These three commodities account for €1,769 mn of the EU's notified AMS in 2000–01. Also, future reforms of the EU wine, fruit and vegetable sectors are not taken into account when calculating the EU's possible reductions of trade-distorting domestic support. These three commodities accounted for €9,603 mn of the EU's notified AMS in 2000–01.

Sources: WTO (2004) and authors' calculations.

3.2.1   Aggregate measure of support

In the case of the AMS, the EU-15 reported that it used €43,654 mn in support of farmers in the WTO amber box, of which the market price support (MPS) accounted for €30,684 mn. For each relevant commodity, the MPS included in the AMS is calculated as the price gap between a fixed world price (average for the period 1986-88) and an administered market price for the concerned commodity, multiplied by the amount produced.[2] In the case of the EU, the administered prices of some commodities that were used in the calculation of the MPS in the 2001 notification by the EU will be or already have been reduced as a result of Agenda 2000, the MTR, and the sugar and olive oil reforms to the CAP. In fact, Agenda 2000 and the sugar reforms abolish the intervention prices for beef and sugar. Until now, these prices have been used in the calculation of the MPS in the EU's notification to the WTO. Under the reforms, the EU introduces a basic price of €2,224 per metric tonne (t) of beef, which will trigger public intervention only when the average market price in a member state or region falls below a safety net level of €1,560/t for two consecutive weeks. This means that the current safety net level in the EU is 70% of the basic price. In the case of sugar the intervention price is replaced by a reference price for sugar that is set at €404.4/t. The new basic and reference prices act as a trigger level for private storage as well as set the level of border protection in the EU. In the case of the olive oil reform, the notified administered price is abolished and replaced by a new trigger price for private storage, which ranges from €1,779/t to €1,524/t for different types of olive oil. It is assumed that these new prices will not be notified as new administered prices given that both the new trigger price for olive oil and the public intervention price for beef are below the fixed world price (the EU's external reference price in its WTO notifications).

These reductions and the abolition of the administered prices, as shown in Table 3.2, will lead to a lowering of the EU's MPS (and hence the

---

[2] Some countries do not use the total amount produced but only the total amount exported.

AMS) by €23,558 mn, given the production level reported in the 2001 notification.[3]

Table 3.2 Agenda 2000 and MTR intervention price reductions

	Notification 2000–01 administered price (€/t)	New administered price (€/t)	External reference price (€/t)
Olive oil	3,837.70	n.a.	2,851.80
Rice	298.40	150.0	143.30
Rye	110.25	n.a.	67.30
Other cereals	110.25	101.3	n.a
Skimmed milk powder	2,055.20	1,747.0	684.70
Butter	3,282.00	2,464.0	943.30
Beef	3,242.00	n.a.	1,729.80
Sugar	631.90	n.a.	193.80

Sources: EU (2003a, 2003b and 2003c) and EU (2006).

Abolishing the administered market prices for beef, sugar and olive oil, combined with reductions in administered prices for cereals and milk, would reduce the notified AMS in 2000–01 to €20,113 mn. This implies that the EU could agree to a reduction of roughly 70% in its final bound AMS commitment (i.e. from €67,159 mn to €20,148 mn).[4]

3.2.2 De minimis

In the case of *de minimis* support, any change in the exempt level would not have any substantial effect on the EU, owing to the fact that the EU's notified *de minimis* support remains quite small. In its 2001 notification to the WTO, the EU's exempt non-product specific and product-specific domestic support amounted to only €538 mn and €40 mn, respectively.

---

[3] This assumes that the EU will use the new basic and reference prices respectively for beef and sugar as the new administered prices in the calculation of MPS in the AMS.

[4] Future reforms to the common market for fruit and vegetables could possibly enable the EU to undertake a larger reduction commitment of the AMS than the 70% presented in this chapter.

According to the text of the July package, it is more likely that a reduction rather than a removal of the exempt *de minimis* support will come out of the Doha negotiations. As such, we conjecture a 50% reduction of the *de minimis* rule.[5] This implies that the *de minimis* rule would be reduced from 5% to 2.5%. In the case of the EU, non-product specific domestic support would then be capped at 2.5% of the total value of agricultural production, which was roughly €6,046 mn in 2001 and was well over the notified number of €538 mn in 2001 (see Table 3.1). As for the product-specific *de minimis* support, the rule of 2.5% implies that the EU would have to reduce the 2001 notification of €40 mn to €23 mn.[6] Therefore, a total of €17 mn would have to be moved back into the calculation of the AMS, which would be the only action the EU needs to take under this conjectured new *de minimis* rule.

3.2.3   Blue box

In the year 2000–01, the EU provided its farmers exempt direct payments under production-limiting programmes in the blue box that amounted to €22,223 mn. These payments were given in the form of land and livestock premiums. Under the MTR reform of the CAP, the majority of these payments will become decoupled from production. Therefore, they can be moved into the undisciplined green box.

Specifically, the implementation of the MTR's single farm payment gives each member state the option to maintain a small part of its direct payments coupled to production. For the EU as a whole, it is expected that 88% of the budgetary transfers in the form of direct payments (i.e. land and livestock premiums, among others) will become part of the single payment. The aggregate ceiling for the single payment in the EU-15 is €33,218 mn for the year 2013, of which around €21 bn was notified as blue box payments in 2001. The MTR will shift a large portion of the blue box payments to the green box, leading to roughly only €4 bn left in the blue box. This amount is

---

[5] In its 2001 WTO notification, the US had over $7 bn in exempt non-product specific and product-specific domestic support. Therefore, a total abolition of the *de minimis* rule does not seem to be a plausible outcome of the Doha round.

[6] This reduction is necessary because the support granted to some commodities in 2001 exceeded 2.5% of their respective production values.

well below the new commitment for the blue box payments (which must not exceed 5% of the total value of agricultural production in the EU, a figure that is in the region of €12,097 mn).

3.2.4   Total trade-distorting domestic support

Given the assumptions about possible reductions in the AMS (amber box), the 50% reduction in the *de minimis* rule and the move of a large share of the blue box payments into the green box, under existing and planned CAP reform programmes the EU should be able to cut its total trade-distorting domestic support from the 2001 notification level of €66,455 mn to a new level of €24,674 mn (see Table 3.1). The overall base commitment of all trade-distorting domestic support of €112,874 mn – from which reductions are to be made – can then be at least reduced by 75%, resulting in a new base commitment level of €28,219 mn. Such a reduction in the total commitment level would leave room for the EU to increase either *de minimis* payments or blue box payments but without the possibility to increase its AMS payments.

3.3   Towards a tiered formula for reducing domestic support in the Doha round

The above analysis has demonstrated that the EU is in a position to institute large cuts to its base commitment level of trade-distorting support within the current reform programmes of the CAP. If indeed the EU adopts such a position, it would have ripple effects on other countries, obliging them to reform their own domestic support programmes.

3.3.1   Proposal on the tiered formula

It can be assumed that the above-proposed cuts (i.e. a 70% cut to the final bound AMS and a 75% cut to the overall base commitments of all trade-distorting domestic support) would put the EU in the first tier of the suggested reform formula. Other countries can then be placed in several different tiers in the conjectured formula according to their current levels of domestic support and their base commitments. Table 3.3 presents this proposal.

Table 3.3a Domestic support base commitments, new commitments and latest WTO notifications – Developed countries

		Currency	AMS	De minimis	Blue box	Total	Production value	Total as a percentage of prod. value	Reduction total/AMS (%)
			(1)	(2)	(3)	(4)=(1+2+3)	(5)	(6)=(4/5)	(7)
Iceland	Base commitment[a)	Mn ISK	130	11	15	156	114	137	75/70
	New commitment[b)		39	2.5%	6	39	–	–	–
	Notif. 2000 [c)		117	0	0	117	–	–	–
Norway	Base commitment	Mn NOK	11,449	1,768	7,880	21,097	17,682	119	75/70
	New commitment		3,435	2.5%	884	5,274	–	–	–
	Notif. 2001		10,700	0	7,240	17,940	–	–	–
Switzerland–Liechtenstein	Base commitment	Mn CHF	4,257	730	365	5,353	7,304	73	75/70
	New commitment		1,277	2.5%	365	1,338	–	–	–
	Notif. 1998		3,273	0	0	3,273	–	–	–
Japan	Base commitment	Bn JPY	3,973	905	452	5,330	9,047	59	75/70
	New commitment		1,192	2.5%	452	1,333	–	–	–
	Notif. 2000		709	32	93	833	–	–	–
EU-15	Base commitment	Mn EUR	67,159	24,194	21,521	112,874	241,943	47	75/70
	New commitment		20,148	2.5%	12,097	28,219	–	–	–
	Notif. 2000		43,654	561	22,223	66,438	–	–	–

Table 3.3a cont.

Canada	Base commitment	Mn CAD	4,301	3,074	1,537	8,912	30,737	29	65/60
	New commitment		1,720	2.5%	1,537	3,119	–	–	–
	Notif. 1999		939	1,102	0	2,041	–	–	–
US	Base commitment	Mn USD	19,103	19,313	9,656	48,072	193,129	25	65/60
	New commitment		7,641	2.5%	9,656	16,825	–	–	–
	Notif. 2001		14,413	7,045	0	21,458	–	–	–
New Zealand	Base commitment	Mn NZD	288	1,338	669	2,296	13,385	17	55/50
	New commitment		144	2.5%	669	1,033	–	–	–
	Notif. 2001		0	0	0	0	–	–	–
Australia	Base commitment	Mn AUD	472	3,493	1,747	5,712	34,934	16	55/50
	New commitment		236	2.5%	1,747	2,570	–	–	–
	Notif. 2002–03		213	20	0	233	–	–	–

Table 3.3b Domestic support base commitments, new commitments and latest WTO notifications – Developing countries

		Currency	AMS	De minimis	Blue box	Total	Production value	Reduction total/AMS (%)
			(1)	(2)	(3)	(4)=(1+2+3)	(5)	(7)
Argentina	Base commitment	USD	75,021,296	–	–	–	–	40/40
	New commitment		45,012,778	5%	–	–	–	–
	Notif. 2000–01		79,599,922	0	0	79,599,922	–	–
Brazil	Base commitment	Thsnd USD	912,105	–	–	–	–	40/40
	New commitment		547,263	5%	–	–	–	–
	Notif. 1997–98		82,820	408,714	0	491,534	–	–
Bulgaria	Base commitment	Mn EUR	520	719	180	1,419	3,594	40/40
	New commitment		312	5%	180	851	–	–
	Notif. 2001		26	9	0	35	–	–
Colombia	Base commitment	Thsnd USD	344,733	–	–	–	–	40/40
	New commitment		206,840	5%	–	–	–	–
	Notif. 1999		6,805	0	0	6,805	–	–

Table 3.3b cont.

Costa Rica	Base	Thsnd USD	15,945	–	–	–	–	40/40
	New commitment		9,567	5%	–	–	–	–
	Notif. 1999		1,595	0	0	1,595	–	–
Israel	Base commitment	Thsnd USD	568,980	632,816	163,620	1,360,001	3,272,391	40/40
	New commitment		341,388	5.0%	163,620	816,000	–	–
	Notif. 2002		248,155	27,131	0	275,286	–	–
Jordan	Base commitment	JOD	1,333,973	111,066,667	27,766,667	140,167,306	555,333,333	40/40
	New commitment		800,384	5.0%	27,766,667	84,100,384	–	–
	Notif. 2002		743,298	10,775,176	–	11,518,474	–	–
Korea	Base commitment	Bn KRW	1,490	6,427	1,607	9,524	32,137	40/40
	New commitment		894	5.0%	1,607	5,715	–	–
	Notif. 2000		1,691	526	0	2,217	–	–
Mexico	Base commitment	Mn MXP '91	25,161	59,164	14,791	99,116	295,821	40/40
	New commitment		15,097	5.0%	14,791	59,470	–	–
	Notif. 1998		3,799	0	0	3799	–	–
Morocco	Base commitment	Mn MAD	685	–	–	–	–	40/40
	New commitment		411	5%	–	–	–	–
	Notif. 2001		300	0	0	300	–	–

Table 3.3b cont.

Papua New Guinea	Base commitment	Mn USD	33	0	-	33	-	40/40
	New commitment		20	5%	-	-	-	-
	Notif. -		-	-	-	-	-	-
South Africa	Base commitment	Mn ZAR	2,015	9,331	2,333	13,679	46,655	40/40
	New commitment		1,209	5%	2,333	8,207	-	-
	Notif. 2002		0	0	0	0	-	-
Taipei China	Base commitment	Mn TWD	14,165	-	-	-	-	40/40
	New commitment		8,499	5%	-	-	-	-
	Notif. -		-	-	-	-	-	-
Thailand	Base commitment	Mn THB	19,028	-	-	-	-	40/40
	New commitment		11,417	5%	-	-	-	-
	Notif. 1998		16,402	0	0	16,402	-	-
Tunisia	Base commitment	Mn TND	59	748	187	994	3,738	40/40
	New commitment		35	5%	187	596	-	-
	Notif. 2000		0	26	0	26	-	-
Venezuela	Base commitment	Thsnd USD	1,130,667	-	-	-	-	40/40
	New commitment		678,400	5%	-	-	-	-
	Notif. 1998		210,578	0	0	210,578	-	-

a) Base commitments on trade-distorting domestic support:

- The AMS (column 1) base level values are taken from the final bound AMS levels agreed in the Uruguay round.
- The permitted *de minimis* (column 2) payments included in the overall base level of trade-distorting domestic support are calculated as a five-tenths percentage of the total value of agricultural production as defined by an average production value in a given reference period (column 5) multiplied by two (total value of non-product- and product-specific *de minimis*).
- Blue box payments (column 3) included in the total base level of all trade-distorting support is the higher of existing blue box payments during the 1995–2000 period or 5% of the value of agricultural production (column 5).
- The total (column 4) value of the overall base level of support is columns (1) + (2) + (3).
- The reference value of agricultural production (column 5) in OECD countries is calculated as the average production value in the period 1999–2002 using values found in the PSE tables. For other countries an average of the reported total value of agricultural production found in the WTO notifications has been used where available.
- In column (6) the total value of the overall base level of all trade-distorting domestic support is calculated as a percentage of the value of agricultural production, with Iceland having the largest percentage value and Australia the lowest among developed countries.
- In column (7) the assumed reduction commitments for the overall base level of trade-distorting domestic support and the AMS is specified, where those developing countries with the highest level of possible trade-distorting domestic support as defined in column (6) make the largest reductions.

b) New commitments:

- The AMS is reduced by 10% less than the overall base level of trade-distorting domestic support.
- The permitted *de minimis* value of domestic support is reduced from a five-tenths percentage of agricultural production value, to 2.5–5%.
- Blue box payments are limited to 5% of the agricultural production value found in column (5).
- The total overall base level of domestic support is reduced by the percentage found in column (7).

c) Notif. refers to the latest notification to the WTO.

Source: Authors' calculations.

In the first tier, four developed countries (Iceland, Norway, Switzerland and Japan) and the EU-15 are grouped together, where a 75% reduction in total trade-distorting domestic support and a 70% cut in the AMS (both of which are from the base commitments) are imposed. These countries have the highest base commitments before any reductions are undertaken, relative to the values of their agricultural production (see Table 3.3). In the next tier are Canada and the US, with the respective proposed reductions of 65% and 60%. The third tier includes New Zealand and Australia, with 55% reductions for both categories. Lastly, it is proposed that developing countries with domestic support commitments reduce both categories by 40%.

The above proposal on reducing the base total commitments and the AMS, along with the new rules on the *de minimis* (the conjectured 2.5% rule for both product and non-product specific support) and the 5% cap on the blue box payments (as in the July package framework), constitutes our best guess as to the possible outcome of the Doha negotiations on domestic support.

3.3.2 How binding are the new commitments?

Using the above proposal and the placement of individual countries in the different tiers, eight countries would have to make reductions to the reported domestic support contained in their latest WTO notifications (see Table 3.3). These of course include those countries in the EU-15 that have not yet fully implemented the MTR reform of the CAP or abolished the intervention price for beef. The other seven countries are Iceland, Norway, Switzerland, the US, Argentina, Korea and Thailand. In this section, we offer a preliminary assessment for the US and discuss very briefly how the other six countries would be able to comply with the proposed cuts.

Total trade-distorting domestic support in the US amounted to $21,458 mn in 2001 (Table 3.4). This figure would have to be reduced to $16,825 mn, given the total base level of $48,701 mn and the proposed 65% reduction commitment. Moreover, both the AMS and the *de-minimis* payments would have to be reduced as these exceeded the proposed new commitment levels. As for the blue box, the US would have no need to take action, since the US did not report anything under the blue box in its 2001 notification to the WTO.

Table 3.4 Proposed domestic support reductions in the US (mn US$)

	AMS	De minimis		Blue box	Total trade-distorting domestic support
		Non-product specific	Product-specific		
Base commitment	19,103	9,656	9,656	9,656	48,072
Notif. 2001	14,413	6,828	217	0	21,458
Of which MPS	5,826	0	0	0	n.a.
Reductions					
MPS reductions					
Milk	-4,483	0	0	0	-4,483
Non-MPS reductions	206	-4,640	-206	4,640	0
New domestic support	10,136	2,188	11	4,640	16,975
New commitment	7,641	4,828	4,828	9,656	16,825

Source: WTO (2004) and authors' calculations.

Can the US get under the proposed new levels of domestic support? To answer this question, we need to review the recent 2002 US Farm Act. The Farm Act maintains the fixed-direct payment (green box) and loan-deficiency payment (amber box) systems but also introduces counter-cyclical payments (CCPs). These new CCPs provide additional payments to producers when market prices fall below a certain level, which is known as the target price. They are expected to be included in the blue box, owing to the fact that these payments are calculated using historical base areas and yields. The CCPs have replaced earlier, ad hoc, crop market-loss assistance payments, which accounted for $4,640 mn of the *de minimis* payments notified by the US in 2001. Therefore, this amount should be moved from the *de minimis* to the blue box, thereby relieving the US of the pressure to drastically reduce its *de minimis* payments. Another area where the US can cut down from the final bound AMS relatively easily and hence the total base commitment is the MPS payments related to the dairy policy. Sumner (2003) noted that the administered prices for the dairy sector could be abolished without hurting US farmers. If this step were taken, it would reduce the AMS reported in 2001 by $4,483 mn (Table 3.4).

After making these changes to the 2001 notification (moving payments into the blue box and abolishing the MPS for milk products), the

US would still not be able to meet the proposed new 65% and 60% reductions. Yet recent USDA long-term baseline projections show that this needs not be the case. These projections are conducted each year in order to forecast the costs of farm programmes for the president's budget. In their latest agricultural baseline projections to 2014 (USDA, 2005), direct government payments to farmers are projected to fall from over $24 bn in 2005 to about $11 bn per year for the period of 2010–14. Towards the end of the projections, direct government payments will largely consist of fixed direct payments under the 2002 Farm Act and conservation payments, which are green box payments. This projection builds on the assumption that government payments fall as rising market prices for programme commodities reduce loan benefits and CCPs to farmers. If these projections are found to be accurate (i.e. if marketing loan gains and loan-deficiency payments are reduced to zero and future blue box payments are drastically reduced), the AMS level notified in 2001 would be reduced by roughly $6,202 mn, thereby enabling the US to comply with the proposed new commitments (Table 3.3).

With regard to Iceland, Norway and Switzerland, their notified AMS levels are mainly comprised of MPS. Therefore, a large reduction in administered prices would be needed to meet their new domestic support commitments as compiled in Table 3.3. In Norway, there would also be a need to reduce blue box payments by reforming its agricultural policy, insofar as to move some of these payments into the green box.

In the cases of Argentina and Korea, domestic support is given to respectively tobacco and rice farmers, in the form of MPS. In Thailand, the majority of domestic support is also given to rice farmers in the form of a paddy-pledging scheme and a soft loan measure. Argentina and Korea would have to reduce their administered prices for tobacco and rice, while Thailand would have to reduce its taxpayer-financed rice support scheme or convert these measures into blue box payments if the proposed reduction scheme was implemented.

3.4 Modelling trade-distorting and non-trade distorting domestic support of the EU

One of the reasons why negotiating domestic support reductions is an integrated part of the Doha agenda is the argument that domestic support programmes – especially amber box measures – encourage over-production of agricultural commodities in developed countries and depress world market prices. In the case of the EU, the budgetary outlays on the CAP programmes also exert great pressure on the EU's common financial scheme. Therefore, one of the ways to create momentum for reducing the spending on these programmes is to reveal the true cost associated with them – an issue that can be analysed with numerical economic models. The modelling issues involved here, however, are far more complicated as compared with the modelling of market access barriers or export subsidies. This is because the functioning of the latter two are more or less uniform across countries, whereas domestic support programmes can vary widely from one country to another.

Recent numerical studies on multilateral reforms of domestic support programmes are largely based on CGE models, among which the GTAP model (accompanied by the global GTAP database) is the most widely used. In the GTAP framework, domestic support measures are modelled as intermediate input subsidies, land and capital-based subsidies, and output subsidies. A sensible modelling approach requires improvements on existing studies in the following areas:

- First, one needs to carefully match the individual domestic-support measures contained in the amber, blue and green boxes with the instruments in the model and relate the various payments to the right instruments.

- Second, reforming domestic measures in many cases calls for changing the association of certain payments from one type of instrument to another type in the model.

- Third, making reductions to domestic support payments often requires maintaining the integrity of the fiscal spending, which usually means that shocking the price wedges is misleading.

- Lastly, the applied shocks in model simulations need to be generated using the differences between the current and the targeted support levels, as opposed to those between the WTO base commitment and the targeted level. This is because there are large differences between the actual spending and the base commitment levels.

For these reasons, modelling the domestic support measures of the EU and their reforms is not a trivial undertaking. Here, we present our modelling of the EU's domestic support programmes in a modified version of the GTAP model (Hertel, 1997), which we have used for many of our studies on CAP reform.[7] The GTAP version 6 database (Dimaranan & McDougall, 2005) is used to carry out the policy simulations.

The domestic support included in the GTAP database originates from the OECD's producer support estimates (PSEs), which have been incorporated into the GTAP database as output subsidies, intermediate input subsidies, land subsidies and capital subsidies. The total value of the PSE included in the GTAP database version 6 (the base year of which is 2001) amounts to €44,785 mn, which includes green, blue and amber box payments but excludes MPS. It is slightly less than the corresponding amount of €47,667 mn notified to the WTO by the EU in 2000–01. The latter includes €22,223 mn in the blue box, €21,845 mn in the green box and €3,600 mn of taxpayer-financed non-exempt direct payments and non-product specific payments in the amber box. The discrepancy is owing to the fact that some payments included in the WTO domestic support notification are not included in the PSE calculation but are found under the OECD's general services support estimates.

3.4.1   Modelling blue box payments of the EU

Blue box payments in the GTAP version 6 database amount to €23,429 mn for the EU-15, €18,031 mn of which are compensatory payments given to farmers based on fixed area and yields (i.e. a hectare premium) and the remaining €5,398 mn of which are compensatory livestock payments based on a fixed number of heads (i.e. a livestock premium). A breakdown of the payments is shown in Table 3.5.

The full implementation of the Agenda 2000 reform of the CAP will increase the blue box payments to €25,358 mn, while the enlargement of EU further increases these payments to €30,587 mn.

---

[7] For a more detailed discussion of the many changes to the standard GTAP model, see the series of working papers published by the Food and Resource Economics Institute on reforms of the CAP and trade liberalisations under the WTO. These papers can be downloaded from www.foi.dk or can be obtained from the authors of this chapter.

Table 3.5 Nominal domestic support moved into the green box (mn €)

	2001	2015	Doha	2015	Doha
		EU-15		EU-10	
Blue box					
Fixed hectare premiums	18,031	17,504	1,961	4,063	0
Fixed livestock premiums	5,398	7,855	2,086	952	0
Total	23,429	25,358	4,047	5,229	0
Amber box					
non-exempt direct payments					
Milk premiums[a]	0	2,936	0	476	0
(Olive oil) [b]	2,469	2,446	622	0	0
Tobacco	964	952	467	0	0
PBF	88	216	0	0	0
Green box					
Extensification premium	914	1,013	0	213	0
Single farm payment	0	0	28,026	0	5,814

[a] New Agenda 2000 compensatory payments given to dairy farmers.

[b] Direct aid given to olive oil farmers, which only seems to be included under the MPS in the amber box.

Source: Authors' calculations.

Those blue box payments based on fixed areas and yields (the hectare premium) are implemented in the model as input subsidies to agricultural land where reform-related crops are grown. In responding to any exogenous shocks to the model, the payments per hectare are fixed but land is allowed to adjust endogenously between reform-related crops as long as the fixed base area is not exceeded. Failing to do so would result in either larger or smaller spending on these programmes than the actual payments. The MTR reform implies that the majority of these payments will become part of the decoupled single farm income payment linked to the utilised agricultural area in each member country.

Similarly, compensatory payments given to livestock (suckler cows and breeding ewes) are modelled as subsidies to agricultural capital, while male animal/steer premiums are modelled as output subsidies to slaughter animals.[8] The implementation of these compensatory payments are modelled by fixing the total EU budgetary expenditure on premiums paid while allowing premiums per cow/ewe and per male animal/steer to adjust endogenously during the projection of the database from 2001 to 2015.

The MTR reform of the CAP converts a large proportion of the livestock premiums into a simple farm income payment in the form of a uniform land-based payment. Therefore, a large proportion of the above-mentioned capital subsidies and output subsidies should be converted to the land-based payment. The decision by the EU to grant each member state the option to keep a small portion of blue box payments coupled to the number of male animals and steers implies that a reduced premium per head needs to be kept as output subsidies, in addition to the uniform land-based payment in some member countries.[9] In these member states, the reduced premium per head is fixed to allow for the budgetary expenditure to adjust to changes in production.

3.4.2   Modelling the green box payments of the EU

Green box payments consist of many different types of support programmes that are supposed to have no or minimal trade-distorting effects. In other words, these payments should be fully decoupled. There are ongoing debates, however, about whether some of them are truly non-distorting. In the GTAP database, all green box domestic-support payments, which are included in the PSE calculations from the OECD, are incorporated into the database as either input or output subsidies. This of course means that green box payments included in the GTAP database are coupled to production to some extent. As modellers and researchers have not reached a consensus on the correct treatment of green box payments in the GTAP, we take the GTAP database as given in this study and focus our

---

[8] These different treatments reflect the fact that suckler cows and breeding ewes are part of the capital used to produce slaughter animals, while male animals and steers are final products sold directly to slaughter houses.

[9] Some member states are also allowed to retain a portion of the hectare premiums coupled to the production of specific commodities.

attention on modelling the decoupling initiatives of the CAP programmes that will result in moving some previous blue and amber box payments into the green box.

Specifically, the MTR reform of the CAP will move a large share of the blue and some amber box payments to the green box. Of the €30,587 mn in the blue box, only €4,047 mn will remain there owing to the MTR (Table 3.5). In addition, direct payments originally classified as amber box payments given to milk, olive oil, tobacco, plant fibres and the extensification premiums in the green box are now all included in the new single farm income payment in the green box.

These decoupled, single farm income payments are incorporated in the model by converting those affected blue and amber box payments in each member country into a uniform hectare payment given to all utilised agricultural area. Therefore, the results found in this chapter represent a decoupling of direct aid from production where no restrictions on the use of land are imposed.[10]

3.4.3 The amber box and the MPS

As discussed earlier, non-exempt domestic support that is classified in the amber box contains both direct payments and the MPS. Direct payments are modelled as various subsidies in the GTAP database and are taken as given as the starting point of our analysis. The decoupling of these payments involves eliminating the coupled payments and increasing the single farm income payments in the form of the uniform hectare payment.

In the case of the EU, another complication is related to the MPS, which comprises a large part of the EU's amber box support and stems from consumer-financed MPS. This is calculated from historical world market prices and administered institutional prices. Because of the historical nature of this measure, it can be quite inaccurate as actual domestic and world market prices may be quite different from their historical references. The inclusion of the MPS in the amber box also hinges

---

[10] The aggregate PSE for the EU is disaggregated by GTAP commodity and EU member state in the version 6 database, whereby the implementation of the proposed movement of amber/blue box payments into the green box shown in Table 3.5 is modelled at the individual EU member state level.

on the existence of an administered price, which might have already been abolished or reduced such that the administered price has no relevance to reality and nor is it used as an active regulatory instrument. If this is the case, modelling the reduction in the MPS means no actual shocks to the model. On the other hand, if reductions in the administered intervention prices are linked to increased compensatory payments, the reductions in the MPS are achieved by reducing the level of border protection (i.e. increased market access and the elimination of export competition – the other two pillars of the Doha agricultural negotiations). In this sense, the MPS reduction is not a pure measure of domestic support but also includes support derived from border measures.

3.5   Simulating the effects of reducing the EU's domestic support

Having outlined how and to what extent the EU could cut its domestic support to accommodate a proposed WTO reform outcome within the reform of the CAP and then described how these reductions are modelled in a CGE model, we are now able to discuss the implications of this reform proposal on the EU and the rest of the world. The impact analysis can be drawn from the differences between a business-as-usual baseline and a Doha reform scenario. So, in this section, we first describe the baseline and the Doha scenario before presenting the numerical results.

3.5.1   The baseline

Like previous rounds of reforms to global trade policy, any multilateral liberalisation following the conclusion of the Doha round will likely take a few years to be implemented. A meaningful evaluation of the anticipated policy changes can be obtained by comparing the liberalisation scenario with a non-liberalisation scenario. Such a non-liberalisation scenario contains projections of the macroeconomy and incorporates the effects of important policy changes other than the exogenous shocks to be analysed. To be consistent with the focus of the study, we construct a non-liberalisation baseline scenario, which features a number of important policy initiatives by the EU, including the Agenda 2000 reform of the CAP, the Everything but Arms initiative and EU enlargement. In addition, the URAA is assumed to be completed in this baseline. Lastly, we apply shocks to GDP, population and total factor productivities to project the world economy to the baseline year of 2015 – a year when we expect the new agricultural agreement to be fully implemented.[11]

3.5.2   The Doha scenario

In the Doha scenario we implement a domestic-support reform programme of the EU (which is largely the MTR reform of the CAP) as discussed in detail in section 3.3, while domestic support programmes in other countries are assumed to be unchanged.[12]

In addition, a stylised interpretation of the July package framework for modalities in reducing market access barriers and export competition measures is also simulated. Specifically, all trade-distorting export competition measures are removed and market access barriers are reduced using the proposed Harbinson multiple-tiered formula on applied tariff equivalents found in the GTAP version 6 database. Unlike the shocks to domestic support reform (which is limited to the EU), the market access and export competition reforms are conducted multilaterally in this scenario. In the analysis of the results, we use the effects of the market access and export competition reforms as a benchmark to gauge the relative magnitude of the impact of the domestic support reform.

3.5.3   Results

Three main results are reported here, including changes in the EU's agricultural production and trade, changes in factor income in agriculture (which we use as a proxy to investigate the effect of the reforms on farmers) and changes in economic welfare. These results are reported in Tables 3.6-3.10. The individual contributions from the EU's domestic support reform and the multilateral market access and export competition reforms to these changes are also provided in the tables.

---

[11] A more detailed discussion of a similar baseline can be found in Jensen & Frandsen (2003b).

[12] We also plan to analyse the reform of domestic support in the US. In the present analysis, agricultural domestic support in the US is fixed exogenously at its nominal value of $32,268 mn as found in the GTAP version 6 database. In the baseline projection, these payments are deflated by 2% a year.

Table 3.6 Change in primary and secondary agricultural production, internal and external trade of the EU, by commodity (%)

| | Change in production | | | | Change in intra-trade | | | | | Change in extra-trade | | | | | | | | |
| | Total | Contributions from | | | Total | | Contributions from | | | Total | | Contributions from | | | Total | | Contributions from | | |
	Prod.	Dom. supp. comp.	Exp. comp.	Mkt access	Trade	Mn $	Dom. supp. comp.	Exp. comp.	Mkt access	Import	Mn $	Dom. supp. comp.	Exp. comp.	Mkt access	Exp.	Mn $	Dom. supp. comp.	Exp. comp.	Mkt access
Paddy rice	-14.7	7.1	-1.4	-20.3	-32.5	-127	6.1	1.1	-39.7	81.4	74	-22.1	-5.7	109.2	83.2	75	39.0	7.5	36.7
Wheat	-3.2	-5.8	-0.6	3.2	0.4	115	-0.9	2.3	-1.0	18.1	76	30.0	-21.5	9.6	-10.8	-132	-20.0	-5.0	14.2
Other grains	-6.4	-2.3	-3.6	-0.5	-3.7	33	-1.6	1.1	-3.1	18.6	105	13.8	-7.0	11.8	-42.4	-317	-13.0	-29.3	0.0
Vegetables, fruit and nuts	1.3	2.4	0.0	-1.1	-3.2	-1,071	1.7	0.5	-5.4	2.7	236	-7.6	-1.7	11.9	19.9	864	8.3	-3.1	14.7
Oilseeds	-14.6	-13.7	0.6	-1.4	-8.3	-38	-11.1	1.2	1.7	12.2	522	16.0	-0.3	-3.5	-26.2	-122	-17.2	1.8	-10.7
Sugar cane and beet	0.1	0.1	-0.1	0.0	0.2	0	0.0	0.1	0.1	-0.2	0	0.0	-0.2	0.0	-0.6	0	0.1	0.2	-0.8
Plant-based fibres	-22.7	-25.0	1.8	0.5	-35.0	-173	-35.1	1.7	-1.6	9.3	118	10.2	-1.6	0.7	-22.1	-204	-26.1	2.3	1.8
Other crops	3.4	3.0	0.4	0.0	-0.5	-260	1.6	0.5	-2.6	-6.2	-425	-9.2	-1.4	4.3	18.0	1,157	11.1	2.1	4.8
Bovine animals	-12	-8.1	-0.6	-3.2	-15.4	-67	-12.1	0.5	-3.8	51.4	166	54.6	-5.4	2.2	-39.0	-363	-46.8	4.0	3.9
Other animals	2.7	1.1	0.4	1.1	1.4	-21	0.6	0.6	0.2	-1.9	-29	-2.8	-1.8	2.6	4.1	93	2.8	0.4	0.9
Raw milk	-0.1	0.2	-0.2	-0.1	2.2	-2	0.6	0.9	0.7	2.2	2	0.6	0.9	0.7	0.0	-5	-0.1	0.1	-0.1
Wool	42.3	33.8	6.4	2.0	43.9	192	36.0	6.3	1.7	-17.5	-134	-13.6	-2.9	-1.1	50.7	484	40.1	8.1	2.6
Bovine meat products	-10.6	-2.9	-3.3	-4.5	-30.1	-1,740	-9.8	0.9	-21.1	56.6	2,216	19.4	-3.1	40.4	-86.1	-1,250	-5.3	-88.6	7.8
Other meat products	2.6	0.4	0.4	1.8	-1.0	-267	2.7	0.4	-4.0	85.6	703	-2.6	-8.3	96.5	49.0	3,063	4.8	4.6	39.7
Vegetable oils and fats	-2.4	-1.3	0.3	-1.4	-13.1	-380	-4.2	0.8	-9.8	39.6	617	9.7	-3.4	33.3	-2.9	-19	-12.9	2.7	7.2
Dairy products	-0.8	0.5	-1.0	-0.4	0.7	-672	0.4	6.6	-6.3	17.3	419	-4.2	-58.2	79.7	-32.7	-1,049	3.2	-57.4	21.6
Processed rice	-24.6	0.5	-4.7	-20.4	-55.9	-314	1.2	1.0	-58.2	80.7	244	-1.5	-1.6	83.8	-64.6	-91	1.7	-68.5	2.2
Sugar	0.3	0.0	0.1	0.1	-4.8	-90	-0.3	5.9	-10.4	-13.9	-362	-0.1	-8.9	-4.8	-54.8	-226	-0.6	-70.3	16.1
Other processed foods	-0.3	-0.1	0.0	-0.2	-2.0	-1,001	-0.1	0.8	-2.7	7.2	1,282	-0.1	-3.2	10.5	3.0	786	-0.1	-3.8	6.9

Source: Authors' calculations.

Agricultural production in the EU

The assumed domestic support reform of the EU has a fundamental redistribution effect on primary agricultural production in the EU, by shifting resources (especially land) from those commodities under the MTR reform to those previously receiving less support. Indeed, outputs of those products that will experience a net drop of domestic support will decrease, including wheat (-5.8%), oilseeds (-13.7%) and bovine animals (-8.1%). On the other hand, the redistribution of support through the uniform land-based payment will boost outputs of those products that previously received less support, such as vegetables and fruit, other crops and other animals, as can be seen in Table 3.6. On aggregate, this reform appears to have little impact on overall agricultural production in the EU, with a mere two-tenths of a percentage decrease from the base production level.

Although the more interesting adjustment lies in the commodity dimension, changes in outputs are also unequal across the EU member states. The total agricultural production of Belgium, Greece, Ireland and the UK is expected to drop and that of other member states (including most of the new ones) is expected to increase. This outcome can largely be explained by the initial production patterns of individual member states, relative to the commodities affected by the domestic reform programme under the MTR.

How large are these output effects, as compared with the impact of multilateral market access and export competition reforms? It appears that the domestic support reform contributes a substantial share of the total decline of outputs of a number of primary agricultural products, including wheat, oilseeds, plant fibres and bovine animals. The redistribution effect of the single farm income payment also leads to output increases of several products such as other crops, other animals, and vegetables and fruit. These output expansions from the domestic support reform contribute significantly to the total output increases found under the scenario. In contrast, the output effects of eliminating export subsidies are generally negative but small, whereas reducing market access barriers has largely negative effects on EU production levels, reflecting the fact that the EU maintains higher average import barriers than many other countries.

Table 3.7 Change in primary and secondary agricultural production, internal and external trade of the EU, by member state (in %)

	Change in production				Change in intra-trade										Change in extra-trade									
	Total	Contributions from			Total		Contributions from			Total		Contributions from			Total		Contributions from			Total		Contributions from		
	Prod.	Dom. supp.	Exp. comp.	Mkt access	Import	Mn $	Dom. supp.	Exp. comp.	Mkt access	Export	Mn $	Dom. supp.	Exp. comp.	Mkt access	Import	Mn $	Dom. supp.	Exp. comp.	Mkt access	Export	Mn $	Dom. supp.	Exp. comp.	Mkt access
Belgium/Lux.	-2.3	-1.4	0.0	-0.9	-4.5	-521	-0.3	1.6	-5.8	-5.0	-654	-2.1	1.5	-4.4	9.7	355	-0.2	-4.3	14.2	0.1	90	-3.9	-7.3	11.3
Denmark	1.9	1.1	-0.1	0.9	-2.2	-91	1.3	1.2	-4.6	-2.0	-223	0.6	3.5	-6.1	9.0	120	-0.5	-4.6	14.1	8.7	485	2.9	-6.0	11.8
Germany	-0.3	0.1	0.0	-0.4	-2.0	-811	0.2	1.7	-3.9	-2.6	-694	0.7	1.7	-5.1	7.5	560	-2.4	-6.9	16.8	3.3	451	0.6	-9.9	12.7
Greece	-3.0	-2.0	-0.2	-0.8	-2.0	-73	0.5	1.7	-4.2	-13.6	-202	-6.9	1.0	-7.7	12.6	158	5.9	-3.4	10.1	-2.4	-13	-3.8	-3.3	4.7
Spain	-1.9	-0.6	0.0	-1.3	-3.9	-301	-0.3	1.7	-5.4	-4.6	-794	0.4	0.7	-5.6	14.2	759	6.7	-3.0	10.4	3.0	191	0.6	-5.3	7.8
France	-0.3	0.1	-0.1	-0.3	-2.5	-533	-0.9	1.2	-2.9	-2.1	-451	0.6	1.7	-4.5	6.1	345	-0.8	-6.1	12.9	-2.6	94	-1.8	-9.5	8.8
Ireland	-10.3	-8.5	-0.2	-1.6	-8.6	-226	-3.2	0.0	-5.4	-14.2	-650	-13.0	3.7	-4.9	21.5	157	9.2	-4.3	16.6	-2.1	7	1.6	-9.2	5.5
Italy	-1.5	-0.1	-0.5	-0.9	-4.5	-602	-0.9	2.7	-6.3	-6.9	-715	1.2	-1.2	-6.9	18.8	994	3.5	-6.8	22.1	7.4	483	1.4	-8.2	14.3
Netherlands	-1.4	-0.3	-0.2	-0.9	-12.9	-1,115	-0.3	1.1	-13.7	-1.7	-703	-0.8	2.5	-3.4	13.7	981	-0.2	-2.9	16.8	4.1	534	0.4	-8.5	12.2
Austria	0.1	0.1	-0.3	0.2	-1.4	-107	0.5	2.1	-4.0	-4.2	-94	0.2	-0.2	-4.1	9.4	65	-3.2	-13.5	26.0	14.0	113	2.1	-9.4	21.3
Portugal	-0.9	0.0	-0.1	-0.8	-4.0	-142	-1.3	1.0	-3.7	-6.8	-65	-2.2	0.3	-4.9	8.2	127	0.4	-2.2	10.0	2.7	18	2.4	-5.2	5.5
Finland	0.0	0.2	-0.3	0.0	-4.7	-71	-0.2	0.9	-5.4	-0.4	-14	0.0	4.5	-4.8	5.8	36	-2.4	-4.0	12.2	-7.3	-10	-0.2	-16.5	9.4
Sweden	3.6	-0.2	-0.3	4.2	-0.4	-49	1.3	1.3	-3.1	-1.7	-34	1.9	1.0	-4.6	11.5	78	-2.8	-7.4	21.7	40.0	447	1.1	-10.2	49.0
UK	-1.6	-0.9	0.0	-0.6	-3.2	-658	0.4	1.9	-5.5	-7.5	-434	-2.2	-0.1	-5.2	8.8	798	2.4	-4.3	10.7	-7.0	-123	-6.3	-7.7	7.0
Cyprus/Malta	-13.4	-0.2	-12.1	-1.1	-9.8	-36	-0.8	0.6	-9.6	0.9	-7	-0.6	7.5	-6.0	7.6	30	0.4	-7.9	15.1	-25.0	-94	0.0	-30.9	6.0
Czech Republic	-1.6	0.9	-2.2	-0.3	-4.0	-113	-1.7	-0.2	-2.1	-2.8	-38	0.1	2.6	-5.5	-0.5	-1	-7.9	-5.5	12.9	-40.2	-149	0.4	-48.3	7.7
Estonia	-3.7	2.0	-4.9	-0.7	-4.3	-21	-1.2	-0.1	-3.1	0.0	-6	3.6	2.5	-6.1	8.6	16	1.8	-5.0	11.9	-7.0	-10	4.1	-17.0	5.9
Hungary	8.9	6.6	-0.6	2.9	-0.7	-24	-0.5	0.3	-0.5	0.7	10	11.0	2.9	-13.2	15.0	57	1.7	-6.8	20.1	27.0	455	1.8	-11.5	36.7
Latvia	-0.4	1.1	-0.3	-1.2	-9.7	-54	-0.9	2.0	-10.7	8.6	7	8.0	-0.3	0.9	24.8	66	-0.1	-3.8	28.7	-1.5	3	-3.4	-5.9	7.8
Lithuania	-1.2	1.4	-2.3	-0.4	-3.8	-29	-3.1	2.7	-3.3	-6.1	-28	-0.3	1.3	-7.1	6.7	17	-3.7	-3.4	13.8	-11.6	-17	-1.2	-21.2	10.8
Poland	-0.1	1.1	-0.9	-0.3	-4.1	-269	-1.5	1.0	-3.7	-1.7	-66	3.6	1.4	-6.6	9.4	109	-4.8	-5.2	19.4	-9.6	-90	1.5	-21.1	10.1
Slovakia	0.3	1.0	-0.5	-0.2	-0.3	-15	-0.8	2.0	-1.5	-0.8	-5	1.6	-0.4	-2.0	-5.5	-7	-10.1	-5.6	10.2	-19.3	-15	-0.2	-23.2	4.0
Slovenia	-6.5	1.8	-9.4	1.1	-3.2	-24	1.0	-1.8	-2.5	-6.7	-26	14.9	4.1	-25.7	6.0	11	-2.4	-9.0	17.5	-26.7	-107	0.7	-46.3	19.0
EU-25	-1.0	-0.2	-0.3	-0.4	-3.9	-5,884	-0.3	1.6	-5.1	-3.9	-5,884	-0.3	1.6	-5.1	10.8	5,830	1.0	-4.9	14.8	1.6	2,745	-0.3	-9.8	11.7

Source: Authors' calculations.

International trade of the EU

Reducing and decoupling domestic support measures will generate uneven effects on the trade patterns and trade volumes of different products. Those commodities subject to the MTR reform programmes will be less competitive in the world market and their exports to external markets will decline. At the same time, imports of these products from the EU's external trading partners will expand. Moreover, increasing external imports, decreasing external exports and shrinking outputs imply that imports within the internal market of the EU will have to be adjusted downwardly, given the normal assumption about consumer behaviour in the EU. For commodities that previously received no or little trade-distorting domestic support, the decoupled support will likely generate opposite effects. A more specific discussion of the results on intra-EU trade and extra-EU trade is presented below (see also Tables 3.6 and 3.7).

- External imports into the EU

    As expected, volumes of external imports into the EU would rise quite significantly for wheat (30%), other grains (13.8%), oilseeds (16%), plant fibres (10.2%) and bovine meats (19.4%) as a result of the EU's domestic support reform. These are the products whose outputs are predicted to decline. In contrast, imports of vegetables and fruit, other crops and other meats would actually decrease. Again, earlier discussion shows that these are the products whose outputs are expected to rise under the MTR reform. On balance, these declines in external imports are more than cancelled out by the increases in the imports of the other products, leading to a 1% gain in total external imports into the EU.

    This redistribution effect along the commodity dimension of the domestic support reform is more evident when compared with the effects of the assumed multilateral market access and export competition reforms contained in the scenario. Specifically, market access reform at the multilateral level increases imports into the EU with very few exceptions and total external imports into the EU rise by nearly 15%, revealing the fact that such reform is more effective in expanding imports into the EU market. Conversely, multilateral reform to export competition almost entirely reduces imports into the EU, with total imports falling by almost 5%. This is because such action turns much of the original exports from the EU into domestic consumption and at the same time pushes up world market prices slightly, making imports more expensive in the EU market, thereby 'crowding out' external imports into the EU. Overall, total external imports into the EU will increase by $5.8 bn owing to the assumed reforms in the three negotiation areas combined.

- External exports from the EU

Changes in external exports from the EU (through its domestic support reform) tend to mirror the corresponding changes in its output and external import patterns. Here, exports of wheat, oilseeds, plant fibres and bovine meats to the external markets of the EU decrease, whereas exports of a few other products actually increase. Total external agricultural exports from the EU will almost remain unchanged. In contrast, greater multilateral market access will mostly create more export opportunities for EU products, resulting in some increases in exports for several products from the EU, which in some cases cancels out the negative effects of the assumed domestic support reform in the EU. For example, the domestic support reform would reduce exports of wheat from the EU by 20%, of which 14.2% would be cancelled out by multilateral market access reform. The impact of eliminated export subsidies would significantly reduce exports of the subsidised products, which reinforces the negative effects caused by the domestic support reform on exports for several products, such as wheat and bovine meats.

Despite drops in exports induced by reducing/decoupling domestic support measures and export subsidies by the EU, increases in the exports of previously unprotected or little-protected products and the opening up of markets elsewhere (caused by the multilateral market access reform assumed in the scenario) will lead to net increases in the EU's external exports by over $2.7 bn. This gain, however, is smaller than the increase in external imports into the EU, thereby leading to a slightly worsened external trade balance in agriculture for the EU.

- Intra-EU trade

  The negative external trade balance implies a reduction of intra-EU trade to some extent. Domestic support reform is quite effective in this regard for commodities under the reform programme. Among primary agricultural products, such reform will reduce the internal trade of oilseeds, plant fibres and bovine animals in the EU, all by over 10%. For processed food products, the internal trade of bovine meats will be reduced by almost 10%. Overall, it is expected that a reduction of three-tenths of a percentage in intra-EU agricultural trade will occur.

  Multilateral market access reforms also generate strong adverse effects on the internal trading of a number of products, especially bovine meats (reduced by 21%). Eliminating export subsidies, however, appears to lead to marginal increases in internal trade, stemming from the redirection of some exports from the external market to the internal EU market. On balance, the three types of reforms will lead to reduced internal trade among the EU member states and the combined drop in intra-EU trade will amount to almost $5.9 bn, most of which represents a displacement by imports from outside the EU.

Changes in factor income in agriculture

The nature of the EU's domestic support reform under the MTR can be further elicited by looking at the total factor payments in primary agriculture. Despite the significant changes in the CAP, the total actual domestic-support spending by the EU will mostly remain unchanged following the MTR reform. As can be seen from Table 3.8, total factor payments in primary agriculture are indeed quite stable for both the original and new member states.

For the original EU-15, these payments only increase by half a percentage point, whereas in the case of the new member states, there is a very small decrease. So, it appears that on aggregate farmers in the EU are almost fully compensated. This point can be made even clearer when comparisons are made against the impact of export competition and market access reforms. In the latter cases, farmers are assumed to receive no compensation. Consequently, total factor payments drop noticeably.

Table 3.8 Change in primary agricultural factor income (in %)

	Total change	Contributions from		
		Domestic support	Export comp.	Market access
Belgium/Lux.	-5.6	-0.9	-2.2	-2.5
Denmark	-2.5	-0.5	-4.3	2.3
Germany	-3.3	1.4	-2.7	-1.9
Greece	-4.1	-2.0	-1.0	-1.1
Spain	-2.7	0.1	-1.2	-1.7
France	-3.7	0.2	-2.6	-1.3
Ireland	-7.1	-0.5	-2.9	-3.7
Italy	-1.1	0.0	-0.6	-0.5
Netherlands	-4.3	-0.8	-2.7	-0.8
Austria	-1.9	0.7	-1.4	-1.1
Portugal	-2.2	0.5	-1.5	-1.2
Finland	-4.7	0.2	-4.2	-0.8
Sweden	-1.4	2.1	-3.6	0.1
UK	-0.9	3.6	-2.3	-2.2
EU-15	-2.9	0.5	-2.0	-1.3
Cyprus/Malta	-3.9	-0.2	-4.4	0.7
Czech Republic	-2.7	0.3	-2.3	-0.6
Estonia	2.2	2.6	-2.0	1.6
Hungary	0.5	0.1	-2.7	3.1
Latvia	-6.3	0.0	-0.3	-6.0
Lithuania	-2.0	0.0	-3.0	1.0
Poland	-2.3	-0.8	-0.9	-0.6
Slovakia	-1.2	0.2	-0.7	-0.7
Slovenia	-4.7	-0.3	-5.3	0.9
EU-10	-1.9	-0.3	-1.6	0.0

Source: Authors' calculations.

Welfare effects

The EU's domestic-support reform programme will no doubt improve the allocation of economic resources among different agricultural sectors and between agriculture and other industries. As such, efficiency gains from this reform are expected. And the gains are larger, the larger the size of the member state and their previously subsidised agricultural sector. Table 3.9 reports these results. For instance, the efficiency gains to Germany, the UK and Spain are all over $400 mn. Overall, the EU-25 will gain in excess of $4 bn from the efficiency improvements arising from its domestic-support reform programmes. When added to the efficiency gains from the multilateral market access and export competition reforms assumed in the scenario, these lead to a total efficiency gain of almost $10 bn.

Table 3.9 Changes in welfare (equivalent variation) (mn US$)

		Contribution from					Contribution from			
	Total EV	Allocative efficiency	Dom. supp.	Export comp.	Mkt. access	Terms of trade	Dom. supp.	Export comp.	Mkt. access	Others
Belgium/Lux.	461	381	149	48	184	203	-33	291	-55	-123
Denmark	519	236	51	134	52	306	-18	246	78	-23
Germany	2,589	1,216	410	376	431	966	230	814	-78	407
Greece	342	340	250	22	67	62	-55	122	-5	-60
Spain	941	857	483	68	307	39	-342	528	-148	46
France	1,951	1,039	355	427	257	795	146	676	-27	117
Ireland	226	464	346	26	92	-238	-45	-114	-78	0
Italy	1,931	1,114	309	161	644	499	-265	784	-21	318
Netherlands	980	877	70	214	595	178	-104	390	-108	-75
Austria	357	163	47	35	80	172	20	136	16	22
Portugal	215	193	91	14	88	26	-8	34	1	-4
Finland	265	113	26	53	33	134	9	124	1	18
Sweden	378	185	46	85	53	147	17	48	82	46
UK	2,601	1,597	761	80	756	920	36	824	59	84
Cyprus/ Malta	-18	56	8	35	14	-24	-4	-28	7	-50
Czech Republic	124	145	80	55	9	22	44	-17	-5	-43
Estonia	33	30	16	12	2	15	-2	10	7	-12
Hungary	353	183	124	41	18	216	46	88	82	-46
Latvia	70	36	17	-2	20	31	4	27	0	3
Lithuania	239	83	64	15	5	151	13	132	6	5
Poland	779	539	391	105	44	289	117	172	-1	-50
Slovakia	175	44	32	9	3	127	10	119	-2	4
Slovenia	9	74	23	60	-9	9	2	-3	9	-73
EU-25	15,519	9,965	4,149	2,073	3,745	5,044	-180	5,403	-179	510

Source: Authors' calculations.

On the other hand, the terms of trade effects of reforming the EU's domestic support programmes are not as uniform, as these are positive for some member states such as Germany, France, the UK and Poland, and negative for Spain, Italy and the Netherlands. These mixed results reflect the fact that reforming domestic support mostly changes the allocation of economic resources among different sectors and its ramifications on the world market prices varies from one product to another. Therefore, member states that experience increases in their net exports (imports) in those products of higher (lower) world market prices will gain from improved terms of trade, while those that suffer losses in their net exports (imports) with lower (higher) world market prices will experience deteriorated terms of trade. The exact effect for individual member states of course depends on their trade structure and the nature of changes to world market prices relative to their trade structure.

The overall terms of trade loss to the EU is about $180 mn, mostly resulting from losses in Spain, Italy and the Netherlands. Combined with the large terms of trade gains from eliminating export subsidies at the multilateral level (and a small loss from market access reform), total terms of trade gains for the EU-25 are expected to be over $5 bn.

These results show that reforms to domestic support in the EU have only minimal influence on the terms of trade of the EU and of its trading partners on the whole. But it does improve the EU's own economic efficiency. The aggregate welfare impact of this reform amounts to just under $4 bn for the EU, which is about one-fourth of the total welfare gains of $15.5 bn to the EU from both the domestic support reform and the assumed multilateral reforms in export competition and market access.

Summary and concluding remarks

Starting with the July package of the WTO agricultural negotiations in 2004, this chapter analyses the implications of a proposal for reforming and reducing domestic support in the EU. Despite the pressure for the EU to undertake large cuts to the final bound AMS agreed in the Uruguay round and to its total trade-distorting support, we find that these cuts can be accommodated within the existing reform programmes of the CAP. Most notably the cuts can be taken under the mid-term review, which decouples a large portion of the EU's blue box payments and some of the amber box payments and makes them eligible for inclusion in the green box. Furthermore, a substantial amount of the market price support that is currently classified and notified as amber box payments has already lost its relevance, such that it can be simply eliminated without any real implications. As such, the 2001 notification of total trade-distorting domestic support of over €66 bn can be reduced to a new level of around €25 bn, implying the feasibility of cutting the EU's overall base commitment of around €113 bn by 75%. In addition, the EU's final bound AMS commitment can be cut by 70%. This proposal of 75% and 70% cuts would effectively restrict the EU from increasing its AMS payments but would leave room for increased *de minimis* and blue box payments.

If the EU takes the lead in implementing this proposal, there would be pressure on other countries to follow suit. We conjecture a reduction formula for all the WTO members, taking into consideration their Uruguay round base commitments and their recent notifications. It is expected that most of these members, including the US, would be able to comply with the reduction formula.

Analysing the effects of such a reduction proposal, however, is not an easy task, owing to several classification, measurement and modelling issues. These issues have more or less been ignored or simplified in many previous, modelling-based studies of reforming/reducing domestic support. As such, this study argues that modelling the reform and reduction of domestic support needs to be conducted in the context of the specific domestic-support programmes of the member countries. Failing to do so would lead to results of little policy relevance. To illustrate these points in the context of reforming the EU's CAP for fulfilling the assumed reduction targets, this study carefully matches the notified domestic support of the EU to the different WTO boxes with the policy variables in a CGE model and database. We then examine how the reform of the CAP changes the nature and size of these payments. Lastly, we re-assign these payments to the right policy variables according to the reform programmes and adjust the fiscal expenditure on the various policy variables as necessary.

Having sorted out the measurement and modelling issues, this study uses the model and database to simulate the likely impact of reforming domestic support in the EU on a unilateral basis. The effect of doing so is benchmarked against a baseline in which no domestic support reform is conducted. In addition, for the purpose of gauging the relative importance of reforming domestic support measures in the Doha negotiations, we also compare the impact of reforms to domestic support with reforms to market access and export competition at the multilateral level.

Several interesting observations from the simulation results are summarised here. First, following the reform of domestic support in the EU, a structural adjustment in EU agriculture and food production is to be expected, with the outputs of wheat, oilseeds, plant fibres and bovine meats dropping significantly. Second, the EU's net export position in the above products is forecasted to deteriorate in response to the reform, while that in other products is expected to improve. The overall size of the EU's agricultural production and trade remain nearly unchanged. These results, which further underscore the redistributive nature of the EU's domestic-support reform programmes, are in great contrast to what can be anticipated from multilateral reforms to market access and export competition. Third, despite substantial allocative efficiency gains accruing to the EU from its domestic support reform, the effect on terms of trade is not that significant. This result stems from the limited and offsetting price effects across different products. Lastly, although gains may be had by other countries that have distinct comparative advantages in those commodities targeted by the EU's domestic-support reforms, overall the welfare effects for other countries are not substantial, as the EU's reforms would have little impact on aggregate trade volume or world market prices.

References

Agra Europe (2005), *Agra Europe Weekly, European Policy News No. 2143*, 11 February.

Baffes, J. and H. de Gorter (2005), *Disciplining Agricultural Support through Decoupling*, Policy Research Working Paper No. 3533, World Bank, Washington, D.C., March.

Dimaranan, B.V. and R.A. McDougall (eds) (2005), "Global Trade, Assistance, and Production: The GTAP 6 Database", Center for Global Trade Analysis, Purdue University, forthcoming.

European Council (2006), Council Regulation (EC) No. 318/2006 of 20 February, on the common organization of the markets in the sugar sector, OJ L 58, 28.2.2006.

————— (2003a), Council Regulation (EC) No. 1787/2003 of 29 September, on the common organization of the market in milk and milk products, OJ L 270, 21.10.03.

————— (2003b), Council Regulation (EC) No. 1785/2003 of 29 September, on the common organization of the market in rice, OJ L 270, 21.10.2003.

————— (2003c), Council Regulation (EC) No. 1784/2003 of 29 September, on the common organization of the market in cereal, OJ L 270, 21.10.2003.

Frandsen, E.F., H.G. Jensen, W. Yu and A. Walter-Jørgensen (2003), "Reform of EU sugar policy: Price cuts versus quota reductions", *European Review of Agricultural Economics*, Vol. 30, pp. 1-26.

Hertel, T.W. (ed.) (1997), *Global Trade Analysis: Modelling and Application*, Cambridge, MA: Cambridge University Press.

Jensen, H.G. and S.E. Frandsen (2003a), *Impact of the Eastern European Accession and the 2003 Reform of the CAP: Consequences for Individual Member Countries*, Working Paper No. 11/2003, Food and Resource Economics Institute, Copenhagen.

————— (2003b), *Implications of EU Accession of Ten New Members: The Copenhagen Agreement*, Working Paper No. 01/2003, Food and Resource Economics Institute, Copenhagen.

Sumner, D.A. (2003), "Implications of the US Farm Bill of 2002 for agricultural trade and trade negotiations", *Australian Journal of Agricultural and Resource Economics*, Vol. 46, No. 3, pp. 99-122.

USDA (2005), USDA Agricultural Baseline Projections to 2014, Office of the Chief Economist, World Agricultural Outlook Board, Baseline Report OCE-2005-1, Washington, D.C., February.

WTO (2004), Notification to the Committee on Agriculture, G/AG/N/EEC/49, Geneva, 1 April.

4. WTO agricultural negotiations
A comparison of recent proposals for market access

Martina Brockmeier, Rainer Klepper and Janine Pelikan

Introduction

A great deal of attention is currently being paid to discussions about the reform of global agricultural trade. In 2000, the World Trade Organisation (WTO) initiated a new round of trade negotiations on agriculture and services. According to the Doha mandate adopted on 14 November 2001, the WTO members committed themselves to substantially improving market access, to reducing (with a view to phasing out) all forms of export subsidies and to significantly reducing trade-distorting domestic support (WTO, 2001).

Several proposals have been delivered by negotiation partners and chairpersons of the agricultural committee on how this Doha mandate can best be achieved. Among them are the revised Harbinson proposal (WTO, 2003a), the US–EU joint text (WTO, 2003b), the Castillo text (WTO, 2003c) and the Derbez text (WTO, 2003d), as well as the Grosser proposal (WTO, 2004a). None of these submitted proposals has lived up to the expectations of a compromise to which all WTO member countries have been able to agree. Only a revised version of the Grosser text (WTO, 2004b), the so-called 'Oshima text', was adopted in July 2004 as part of the Doha Work Programme (WTO, 2004c). Yet the adopted Doha Work Programme is very vague. It contains almost exclusively qualitative information about tariff cuts and the abolition of export subsidies, but does not make any concrete statements regarding the time horizon or magnitude of the tariff reductions. The latest attempt to bridge the gap between the diverging interests of the WTO members involves new proposals by the EU, the US, the G-20,[1] the G-10[2] and the African, Caribbean and Pacific (ACP) countries,[3] which were delivered in October 2005. But the perspectives of the WTO member countries remained too diverse; it was evident before the last Ministerial Conference in Hong Kong in December 2005 that the newest proposals would meet the same fate as their predecessors.

One of the most difficult points to agree upon in the WTO agricultural negotiations is certainly the expansion of market access. Here, the suggestions already made range from a repetition of the Uruguay round formula to an ambitious cut of 90% for tariffs above 60% in the initial situation. Given these widespread proposals, it is not surprising that WTO member countries have only made marginal progress in the pillar of market access. Until now, WTO members could only agree to a standardised conversion of specific tariffs in *ad valorem* tariff equivalents (AVEs). Additionally, the use of a tiered harmonisation formula with four bands was accepted by all WTO member countries (WTO, 2004c). Yet most other questions dealing with market access remain unresolved. The level of tariff reductions and type of tiered harmonised formulas, the coverage of tariff bands as well as the tariff band flexibility are probably the most prominent issues. But the handling of sensitive products and tariff quotas also impair the already-hardened trade positions of WTO member countries.

Against this background, it is interesting to analyse whether the outcomes of the most recent proposals for market access are so different that they justify the halt of the WTO negotiations. For this reason, section 4.1 briefly illuminates the market access aspects of the latest proposals submitted by the EU, the US, the G-20 and the G-10. In section 4.2, the Global Trade Analysis Project (GTAP) framework is introduced together

---

[1] At the time of the proposal, the G-20 included Argentina, Bolivia, Brazil, Chile, China, Cuba, Ecuador, Egypt, India, Mexico, Nigeria, Pakistan, Paraguay, Philippines, South Africa, Thailand, Venezuela and Zimbabwe.

[2] The G-10 includes Bulgaria, Iceland, Israel, Japan, the Republic of Korea, Liechtenstein, Mauritius, Norway, Switzerland and Chinese Taipei.

[3] The ACP countries also submitted a proposal. It does not include quantitative suggestions for tariff cuts and thus it is not considered here.

with the theoretical extensions. Thereafter, model design and experiments are discussed in section 4.3, followed by results (in section 4.4) and conclusions.

4.1   Overview of the most recent proposals for market access

Numerous proposals concerning market access were made by different negotiation groups in advance of the WTO Ministerial Conference in Hong Kong in December 2005. The proposals of the EU, the US, the G-20 and the G-10 deliver quantitative information on the tariff cuts by taking the four bands and the use of a tiered harmonisation formula into account (see Table 4.1).

Table 4.1 Proposals for market access by the EU, the US, the G-20 and the G-10

EU proposal		US proposal			G-20 proposal		G-10 proposal	
Tariff rate (%)	Tariff cut (%)	Tariff rate (%)	Tariff cut (%) initial final		Tariff rate (%)	Tariff cut (%)	Tariff rate (%)	Tariff cut (%)
Developed countries								
> 90	60	> 60	85	90	>75	75	> 70	45
> 60 ≤ 90	50	> 40 ≤ 60	75	85	>50 ≤ 75	65	> 50 ≤ 70	37
> 30 ≤ 60	45	> 20 ≤ 40	65	75	>20 ≤ 50	55	> 20 ≤ 50	31
0 ≤ 30	35	0 ≤ 20	55	65	0 ≤ 20	45	0 ≤ 20	27
Cap: 100%		Cap: 75%			Cap: 100%		-	
Developing countries								
> 130	40	> 60	n.a.	n.a.	>130	40	> 100	n.a.
> 80 ≤ 130	35	> 40 ≤ 60	n.a.	n.a.	>80 ≤ 130	35	> 70 ≤ 100	n.a.
> 30 ≤ 80	30	> 20 ≤ 40	n.a.	n.a.	>30 ≤ 80	30	> 30 ≤ 70	n.a.
0 ≤ 30	25	0 ≤ 20	n.a.	n.a.	0 ≤ 30	25	0 ≤ 30	n.a.
Cap: 150%		n.a.			Cap: 150%		-	

Sources: USTR (2005), G-10 (2005), G-20 (2005) and FAPRI (2005).

Table 4.1 reveals that expectations about the impact of tariff reductions on opening markets in the WTO member countries are very widespread. They range from tariff cuts as high as 90% for tariff rates above 60% in the US proposal, to the far lower tariff cuts of 45% for tariff rates above 70% in the G-10 proposal. It can also be seen that the EU proposal asks for more moderate tariff cuts, while the G-20 suggest tariff cuts that are between those in the EU and US proposals. The cap on tariffs in the US proposal asks for an ambitious reduction of all tariffs below 75% in developed countries. By contrast, the EU and the G-20 would be satisfied if the final tariff rates of developed countries do not exceed 100%.

Meanwhile, the G-10 does not want to apply a cap at all. Table 4.1 also reveals that the proposals do not always offer complete information on tariff cuts in developing countries (i.e. the US and G-10 proposals).

Although the use of a tiered formula with four bands has already been decided, Table 4.1 shows the leeway that exists in implementing this formula. The imposition of a tiered formula with linear cuts between the bands, however, implies the problem of discontinuity, which results in a change of the ordering of tariffs. From the political–economy perspective, such discontinuities would create political resistance from agents in sectors that are just above the transition points (Anderson & Martin, 2005, p. 16). Also, developing countries (such as the Dominican Republic) with fixed bound tariffs at one specific level can be strongly affected by the problem of discontinuities. One possibility for avoiding this problem is the implementation of a progressive tiered formula as proposed by Canada in May 2005.[4] Instead of applying a single cut to the entire tariff line, different cuts are applied to different portions of the same tariff. Because of smaller cuts in the lower portions of the tariff, this formula cuts high tariffs by less than a linear tiered formula in absolute terms. Yet this potential tariff-cutting option is not discussed in the recent proposals on how to open market access. The G-10 proposal avoids the problem of discontinuity by adjusting a limited range of tariffs surrounding the thresholds (see also section 4.3).

4.2   GTAP framework

4.2.1   Extension of the GTAP model

The analyses in this chapter are based on the comparative static, multi-regional general equilibrium GTAP model. This model provides an elaborate representation of the economy, including the linkages between the farming, agribusiness, industry and service sectors of the economy. The use of the non-homothetic, constant difference of elasticity functional form to handle private household preferences, the explicit treatment of international trade and transport margins and a global banking sector that links global savings and consumption are innovative in the GTAP. Trade is

---

[4] For further details, see the website link (http://www.tradeobservatory.org/library.cfm?refid=72991).

represented by bilateral trade matrices based on the Armington assumption. Further features of the standard model are perfect competition in all markets as well as a profit- and utility-maximising behaviour on the part of producers and consumers. All policy interventions are represented by price wedges. The framework of the standard GTAP model is well documented in the GTAP book (Hertel, 1997) and is available on the Internet.[5]

Agricultural policy instruments are represented through price wedges in the standard GTAP model. Therefore, the standard GTAP model is extended with an explicit modelling of the instruments related to the EU's mid-term review (MTR). Following the approach of Frandsen, Gersfeld & Jensen (2002), we introduce an additional land subsidy rate into the model that is equalised across all sectors entitled to direct payments.[6] With the implementation of the MTR, the existing domestic support measures are converted into a region-specific, fully decoupled, land-area payment, while budgetary outlays for total domestic support are held constant. We deliberately did not model the EU sugar policy, as this would require resources that go far beyond the scope of this study (see Brockmeier, Sommer & Thomsen, 2005).

The EU budget is introduced in the GTAP model using a social accounting matrix (SAM). This SAM not only covers the expenditures and revenues of existing agents (producers, government, private households, etc.), but also of the European Agricultural Guidance and Guarantee Fund (EAGGF). This EU budget receives 75% of the import duties for agricultural and non-agricultural products from producers, private households, the government and the capital account. Additional revenues result from an endogenously calculated GDP-related tax that flows from regional households to the EU budget. Here, all EU member countries face an equal GDP tax rate. The revenues of the EU budget are used to cover agricultural output and export subsidies as well as direct payments. In contrast to these product-specific instruments, expenditures on structural policies are not covered by the EU budget module. Because of their characteristics and

---

[5] For further details, see the GTAP website (http://www.gtap.agecon.purdue.edu/products/gtap_book/default.asp).

[6] We are grateful to Hans Jensen for his support in implementing the element of decoupling.

specific aims, structural funds cannot be allocated to certain commodities – strongly hampering their implementation in a product-specific model such as the GTAP.

Obviously, the revenues of the EU budget from one member state are not identical to the expenditures of the EU budget on the same member state. A comparison of the revenues and expenditures of each member state therefore shows the net transfer that takes place within the EU financial system. Analogous to capital transfer, the net transfer within the EU is part of the current account balance that makes up the difference between the exports and imports of goods and services. Yet the sum of net transfers of all member states equals zero, since the EU budget is balanced through the endogenous GDP tax rate.

In the standard GTAP model, EAGGF revenues and expenditures are organised through the regional household. All components of the EU budget are therefore introduced with the help of dummy variables allowing an easy shift from regional households to the EU budget and vice versa. Consequently, a preliminary simulation is employed to move the GTAP database from the initial situation without the EU budget to a new equilibrium, wherein the EU budget is in charge of the EAGGF (Brockmeier, 2003, pp. 100-12).

Alongside changes in the political environment of an economy, macroeconomic developments such as technical progress are of great importance to economic growth. In order to take these changes into account, corresponding trends are incorporated into the analyses at hand. For this purpose, we include in the extended GTAP model exogenous projections for regional GDP and factor endowment based on data from Walmsley et al. (2000). In the simulations, technical progress is generated endogenously by the model, enabling the projected growth pattern.

4.2.2 Extension of the GTAP database

The most recent GTAP database (version 6.04) includes applied tariffs, which are based on the Market Access Map (MAcMap). The source files of MAcMap come from the UNCTAD's Trade Analysis and Information System (TRAINS), the WTO and the Agricultural Market Access (AMAD) database. The applied rates of the newest GTAP database therefore take preferences, AVEs and tariff-rate quotas (TRQs) into account. Information on preferences in MAcMap is taken from the TRAINS database and is augmented with data from national sources. AVEs are calculated on the basis of the median unit value of worldwide exporters, using an average flow of the years 2000–03. Finally, TRQs are taken into account by utilising the filled rate from the AMAD database. If the filled rate is less than 90%, the in-quota tariff is used. The out-of-quota rate is employed if the filled rate is higher than 100%. If the filled rate is higher than 90%, but lower than 100%, a simple average of the in-quota and out-of-quota rate is applied (Bouët et al., 2004).

WTO negotiations take place at a much higher, disaggregated level of tariff lines. To be as close as possible to this negotiation process we implement all tariff cuts at the HS 6 tariff-line level (see 4.3.2). For this purpose, we supplement the GTAP database with a tariff module that includes bound and applied rates at the HS 6 tariff-line level. The applied and bound rates at the HS 6 tariff-line level are taken from the MAcMap database.

4.3   Empirical analysis

The simulations are based on the GTAP database version 6.04, with 2001 as the base year. The database consists of bilateral trade, transport and protection matrices that link 57 sectors in 87 countries or regions. In order to keep the calculation effort within a reasonable scope, the database is aggregated into 23 regions and 19 sectors (see Tables A4.1 and A4.2 in the appendix). The regional sets are put together with regard to geographical nearness, developmental status, or membership in a certain regional agreement. With regard to the sectoral aggregation, it was important to distinguish between primary and processed agricultural-production sectors.

4.3.1   Simulations

Before the actual simulations are carried out, it is necessary to conduct some pre-simulations to implement the extended model structure and to update the protection rates (see Figure 4.1 and Table A4.3 in the appendix).

This pre-simulation includes instruments used by the common agricultural policy (CAP) and the common budget of the EU. Based on the results of the pre-simulation, a base run is conducted that represents a projection of the exogenous variables of population, GDP and factor endowment up to the year 2014. Additionally, Agenda 2000, EU enlargement and the Everything but Arms (EBA) agreement as well as the

MTR are implemented in 2004, 2010 and 2014, respectively (for details see Table A4.3 in the appendix). The base run only considers political intervention in the EU-15 and in the candidate countries. Developments in other regions, such as the Farm Bill in the US or China's WTO accession, are not yet taken into account. Parallel to the base run, a scenario is implemented as well. It takes account of the same projections and policy shocks (Agenda 2000, EU enlargement, the EBA agreement and the MTR). Additionally, it implements the negotiations of the WTO round in the period from 2010 to 2014. Here, tariff cuts to open market access as proposed by the EU, the US, the G-20 and the G-10 are employed as shown in Table 4.2. The US and the G-10 proposals do not provide information about the magnitude of tariff cuts or the cap for developing countries. Analogous to a study by FAPRI (2005), these cuts are assumed to be two-thirds of the cuts for developed countries. The effects of the WTO round are obtained by comparing the results of the base run with those for the scenario for 2014.

Figure 4.1 Base run and simulations

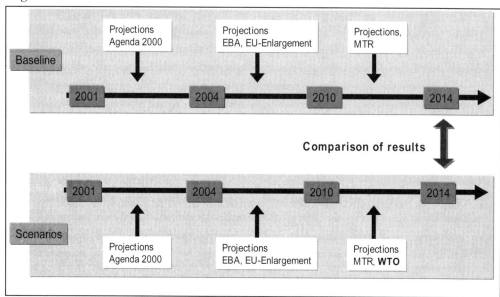

Table 4.2 Proposed tariff cuts by the EU, the US, the G-20 and the G-10 used in the simulations

EU proposal		US proposal			G-20 proposal		G-10 proposal	
Tariff rate (%)	Tariff cut (%)	Tariff rate (%)	Tariff cut (%) initial	final	Tariff rate (%)	Tariff cut (%)	Tariff rate (%)	Tariff cut (%)
Developed countries								
> 90	60	> 60	85	90	>75	75	> 70	45
> 60 ≤ 90	50	> 40 ≤ 60	75	85	>50 ≤ 75	65	> 50 ≤ 70	37
> 30 ≤ 60	45	> 20 ≤ 40	65	75	>20 ≤ 50	55	> 20 ≤ 50	31
0 ≤ 30	35	0 ≤ 20	55	65	0 ≤ 20	45	0 ≤ 20	27
Cap: 100%		Cap: 75%			Cap: 100%		-	
Developing countries[a]								
> 130	40	> 60	56.7	60	>130	40	> 100	30
> 80 ≤ 130	35	> 40 ≤ 60	50	56.7	>80 ≤ 130	35	> 70 ≤ 100	24.7
> 30 ≤ 80	30	> 20 ≤ 40	43.3	50	>30 ≤ 80	30	> 30 ≤ 70	20.7
0 ≤ 30	25	0 ≤ 20	36.7	43.3	0 ≤ 30	25	0 ≤ 30	18
Cap: 150%		Cap: 112.5%			Cap: 150%		-	

a) The US and the G-10 proposals give no information about the magnitude of tariff cuts or the cap for developing countries. Analogous to a study by FAPRI (2005) these cuts are assumed to be two-thirds of the cuts for developed countries.

Sources: USTR (2005), G-10 (2005), G-20 (2005) and FAPRI (2005).

4.3.2 Calculations of tariff cuts

The negotiations of the WTO are based on bound rates, while the economic effect of a tariff cut clearly depends on the applied rate. Therefore, our calculations of tariff cuts take both kinds of tariff rates into account. The difference between bound and applied duties is called 'water in the tariffs'.[7] A reduction of the bound rate does not result in a trade effect if the reduced bound rate is above the applied rate (see Figure 4.2, parts 1.1 and 1.2), e.g.

---

[7] There is disagreement over the definition of the term 'water in the tariffs' in the literature. For example, Martin & Wang (2004) define water in the tariffs as any gap between the applied rate and the actual rate of protection, where the actual rate is lower. Additionally, the term water in the tariffs is not equivalent to the term 'binding overhang', which defines the difference between the bound and the most-favoured nation rate (Francois & Martin, 2003).

water in the tariff still exists after the tariff cut, so that imports are unchanged. But there will be a trade effect if tariff cuts exceed the water in the tariffs (Figure 4.2, part 1.3).[8]

Figure 4.2 Bound rates, applied rates and water in the tariffs

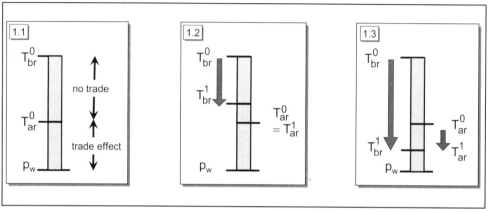

Notes: T = tariffs, br = bound rates, ar = applied rates, p_w = world market price

Accordingly, tariff cuts in the proposals of the EU, the US, the G-20 and the G-10 are calculated at the HS 6 tariff-line level based on the following equations:

$$T_{br}^1 = T_{br}^0 \cdot (1 - \frac{y_{br}}{100})$$
(1)

where:

T = tariff rate

y = tariff cut in %

subscript br/ar = bound/applied rate

superscript 0/1 = initial/final situation.

---

[8] Because of the lack of available information we do not take into account the effective protection. It should be stressed, however, that an implemented tariff cut will not result in a trade effect if it leaves the applied rate above the effective protection. The effective protection is defined as the amount by which the prevailing internal price exceeds the world market price before tariffs (Podbury & Roberts, 2003, p. 5).

If T_{br}^1 is higher than or equal to T_{ar}^0, no tariff cuts will be implemented. If T_{br}^1 is smaller than T_{ar}^0, the tariff cut to achieve $T_{br}^1 = T_{ar}^1$ will be implemented according to equation (2):

$$T_{br}^1 = T_{ar}^1 = T_{ar}^0 \cdot \left(1 + \left(\frac{T_{br}^1 - T_{ar}^0}{T_{ar}^0} \right) \right) \tag{2}$$

Water in the tariffs will lead to country-specific reduction commitments. Owing to the option to bind the ceiling, developing countries were allowed to implement the tariff binding without reference to former protection levels. As a result, the bound tariffs in developing countries are much higher than in developed countries (Anderson & Martin, 2005, p. 14). Therefore, developing countries may experience an implicit preferential treatment that may be additional to the already granted special and differential treatment.

The new applied rate (T_{ar}^1) is aggregated to the GTAP level using import trade weights. This is done with the help of source generic world-import values from the UN's Commodity Trade Statistics Database (COMTRADE) from the year 2001, excluding intra-EU trade. Import weighting is the most commonly used aggregation scheme, which is also utilised to aggregate the applied rates included in the GTAP database version 6.04. Advantageously, trade weights take the relative importance of trade flows into account. Furthermore, the welfare implications are better addressed with this method. By contrast, the import-weighted aggregation scheme leads to an endogenous bias, as the weight for each individual tariff decreases with an increase in the tariff. Accordingly, prohibitive tariffs that impede market access and thereby reduce the trade volumes to zero are not taken into account by import weighting. Trade barriers are thus underestimated with this method.[9]

Finally, we calculate the shocks at the GTAP level that are necessary to reduce the initial applied rate of the GTAP database (T_{ar}^0) to the new

---

[9] In contrast to this study, Walkenhorst & Dihel (2003) used simple averages for the tariff aggregation to avoid biases from the interdependence of tariff levels and trade flows. The simple non-weighted average, however, does not take the relative importance of particular tariffs into account.

applied rates calculated with the help of the tariff module, for all the WTO proposals on market access under consideration (see Table 4.2).

4.4   Results

This section discusses the results of four experiments analysing the implementation of the tariff cuts proposed by the EU, the US, the G-20 and the G-10 in advance of the Ministerial Conference in Hong Kong in 2005. The results are presented in millions of US dollars for the year 2001 of the GTAP database and are obtained using version 9 of the software GEMPACK (Harrison, Horridge & Pearson, 1996). As a macroeconomic closure we adopt a fixed trade balance.

In this section, we mainly focus on the EU trade balance. The appendix provides detailed results for production output (in Table A4.5) on a disaggregated country level. Changes in production output are primarily induced by changes in the trade regime. The output results therefore show a pattern that is similar to the changes in trade balances and is only discussed in a rudimentary way.

4.4.1   Impact on the EU trade balance

Table 4.3 reports the changes in the regional trade balance[10] by commodity resulting from the implementation of each of the four proposals. Examination of the entries in Table 4.3 shows that the changes in the EU-27 are negative in almost every important sector in all four simulations. These negative changes are particularly pronounced in the highly protected beef and milk sectors, where imports rise relative to exports under the EU proposal by more than $-8.05 bn and $-8.23 bn respectively. Table 4.3 also reveals that the change in the EU trade balance for milk does not significantly vary between the simulations. Notably, the EU trade balance for beef differs by around $7.48 bn when the US proposal ($-14.46 bn) is implemented instead of the G-10 proposal ($-6.98 bn).

---

[10] The change in the trade balance represents the change in the value of fob exports minus the value of cif imports.

Table 4.3 Changes in trade balances owing to the implementation of proposals by the EU, the US, the G-20 and the G-10 (mn US$)

	EU27	USA	Japan	Oceania	WTO IC	Brazil	India	ACP	LDC	WTO DC	ROW
EU proposal											
cereals	-388	3284	37	3	290	743	-59	-51	83	-4307	-58
oilseeds	181	1363	137	-133	-785	2461	21	-86	24	-3368	-150
paddy rice	-4624	-92	-4588	46	-81	-195	357	-41	-1126	9288	-128
vegetables and fruits	-214	-939	-102	-102	229	-129	-172	-130	-117	1466	-117
cattle	176	-632	27	-323	205	-12	1	-36	-18	730	-37
other animal	-130	-1098	26	-200	33	-55	28	-41	-46	1674	-137
beef	-8053	1857	-1189	648	117	3756	-1172	197	72	1554	2246
other meat	-2168	1352	-3351	-142	-907	-2078	10	-61	183	7182	-70
milk	-8232	992	-406	2162	630	24	364	416	588	2838	1037
sugar	-1708	-185	-383	353	-8	106	-14	887	-955	1947	43
other food	-7127	-4448	-6146	181	-592	-1393	-1191	-123	-608	21116	-888
other primary	3397	1619	595	-43	473	-221	-890	307	-1474	-5321	-826
manufactures	6542	-15515	15765	-2243	-2251	-1390	3176	-1283	4707	-23519	1944
services	22305	12449	-421	-203	2674	-1612	-540	59	-1304	-11307	-2750
US proposal											
cereals	-39	3505	-28	39	724	848	-133	-296	138	-5596	140
oilseeds	289	1786	155	-155	-752	1894	52	-159	105	-3429	-131
paddy rice	-5797	115	-4003	77	-99	-259	617	-74	-1346	9530	-130
vegetables and fruits	-401	-960	-213	-103	572	-276	-994	-120	-96	1828	-55
cattle	332	-901	49	-437	362	-50	3	-51	-14	880	-38
other animal	-81	-957	95	-196	110	-142	75	-46	-48	1342	-117
beef	-14466	2987	-2882	1291	781	12455	-4830	297	86	1639	2255
other meat	2985	3783	-6901	-303	-679	-3435	-187	-211	266	4235	-82
milk	-7863	1833	-1646	4571	-1922	-70	324	183	643	2189	1793
sugar	-4499	-934	-1439	1044	3	429	-5	3341	-2337	3977	102
other food	-7234	-4139	-7044	411	-490	-1416	-1746	-402	-445	21160	-737
other primary	3876	1148	1290	-590	613	-1433	-143	-30	-1140	-4924	-859
manufactures	8461	-19493	21773	-4665	-2277	-5948	6817	-2290	5314	-22441	637
services	24417	12220	794	-976	3071	-2590	27	-108	-1132	-10353	-2665

1) For the composition of regions and sectors see Table A1 and A2 in the appendix. 2) WTO IC: other developing WTO member countries; ACP: African, Caribbean and Pacific countries; LDC: Least Developed countries; WTO DC: other developing WTO member countries; ROW: Rest of the world

Source: Own calculations.

Table 4.3 cont.

	EU27	USA	Japan	Oceania	WTO IC	Brazil	India	ACP	LDC	WTO DC	ROW
G-10 proposal											
cereals	-308	2369	53	163	640	-23	0	64	130	-3224	16
oilseeds	90	78	113	-129	-829	2948	11	-78	-49	-2300	-167
paddy rice	-883	-478	-7444	-63	-20	-173	-60	-46	-1043	9198	-132
vegetables and fruits	-260	-363	-6	-68	-1	-91	-26	-108	-107	968	-129
cattle	145	-386	8	-268	62	-6	0	-30	-16	597	-35
other animal	-107	-861	5	-173	-59	-33	18	-37	-43	1467	-127
beef	-6976	1825	-844	523	84	1797	-75	125	69	1313	2249
other meat	-3385	1446	-2530	-88	-399	-1604	302	-32	166	6225	-65
milk	-8330	884	-17	1878	887	44	388	428	574	2767	975
sugar	-1302	-5	-114	190	-15	85	38	423	-548	1360	29
other food	-6614	-4652	-3860	-319	-601	-1264	-857	-11	-572	18541	-639
other primary	2796	2435	195	73	296	69	-1096	365	-1605	-5106	-843
manufactures	4319	-14607	15024	-1692	-2424	-340	2040	-1122	4433	-21462	1809
services	20763	12319	-581	-26	2418	-1407	-763	70	-1378	-10377	-2838
G-20 proposal											
cereals	-321	3233	54	11	429	602	-57	-124	95	-4320	-42
oilseeds	219	1463	152	-138	-773	2258	29	-107	40	-3330	-145
paddy rice	-4651	-94	-4580	45	-79	-223	376	-57	-1206	9419	-130
vegetables and fruits	-402	-1056	-138	-148	230	-175	-117	-142	-104	1770	-113
cattle	238	-700	31	-324	256	-23	2	45	-15	707	-37
other animal	-118	-1085	41	-207	57	-86	41	-45	-43	1632	-136
beef	-10301	2129	-1578	687	183	7153	-2841	282	78	1910	2245
other meat	-968	2848	-5583	-177	-827	-2560	-89	-110	213	7127	-55
milk	-8264	1072	-820	2735	-106	8	376	399	606	3126	1198
sugar	-2607	-407	-556	469	-7	205	-18	1725	-1549	2662	66
other food	-7317	-4222	-6737	332	-633	-1499	-1114	-177	-533	21445	-894
other primary	3595	1546	873	-155	535	-656	-688	171	-1344	-5370	-825
manufactures	8003	-16928	18800	-2753	-2039	-3030	4351	-1731	5006	-25455	1716
services	22854	12208	41	-371	2792	-1968	-338	-19	-1238	-11328	-2740

1) For the composition of regions and sectors see Table A1 and A2 in the appendix. 2) WTO IC: other developing WTO member countries; ACP: African, Caribbean and Pacific countries; LDC: Least Developed countries; WTO DC: other developing WTO member countries; ROW: Rest of the world
Source: Own calculations.

What causes these different results? In the following discussion, the results are decomposed and examined more closely. Here, the total changes are disaggregated into parts, the so-called 'subtotals', which are attributable to changes in individual exogenous variables, e.g. policy instruments. The decomposition of the total effect into subtotals thereby allows the identification of changes that govern the results. The decomposition is based on the changes in policy instruments (e.g. import tariffs or export subsidies) that are applied to bilateral trade flows. Thus, the policy instrument, as well as the source and the destination of the trade flow subject to the policy instrument, can be identified (Harrison, Horridge & Pearson, 1999).

Such a decomposition of the EU trade balances for selected products is presented in Table 4.4. It shows that in the case of beef, the results are clearly driven by two main effects. The first effect involves a deterioration of the EU trade balance between $-0.96 bn (G-10 proposal) and $-8.41 bn (US proposal), which is induced by the cuts of EU tariffs on agricultural products that are imported from third countries.

The second effect stems from the increase in EU beef imports relative to beef exports by $-4.54 bn (G-10 proposal) and $-4.77 bn (G-20 proposal), owing to the abolition of export subsidies. While the latter remains more or less unchanged between the simulations, it can obviously be seen from Table 4.4 that the EU beef sector reacts very sensitively to EU agricultural tariff cuts. Additionally, Table 4.4 reveals the minor importance of the opening of third countries' beef markets to the exports of highly-protected EU beef.

In contrast, the results for the EU trade balance in processed milk trade are almost completely dominated by the abolition of EU export subsidies (see Table 4.4). This effect varies only slightly between the simulations and comprises a relative increase of EU milk imports from around $-10.16 bn (G-10 proposal) to $-11.73 bn (US proposal). Milk exports are also subsidised by third countries. In this regard, the abolition of export subsidies results in an expected positive effect for the EU trade balance for milk. The effect, however, at $0.31 bn, is negligibly small compared with that from dismantling EU export subsidies. Also, all other repercussions, particularly those arising from the cut to EU import tariffs, are insignificant. Even the high cuts to agricultural tariffs in the US proposal evoke only smaller changes in the EU trade balance for milk. Here, the better access to third countries' milk markets (to the extent of $6.98 bn) is almost offset by the effect of EU agricultural tariff cuts ($-2.16 bn) and of diverting EU milk exports, which results from better market access among third countries ($-1.62 bn).

Table 4.4 Decomposition of the changes in the EU trade balances for selected products under proposals from the EU, the US, the G-20 and the G-10 (mn US$) [a), b)]

| Commodity | Proposal | Tariffs of agricultural product from | | | Tariffs of non-agricultural products | Export subsidies from | | Sum |
		TC to EU	EU to TC	TC to TC		EU to TC	TC to all	
Cereals	EU	134	586	-346	53	-802	-13	-388
	US	101	1137	-506	53	-806	-17	-39
	G-10	87	456	-66	53	-824	-13	-308
	G-20	171	589	-319	52	-800	-14	-321
Sugar	EU	-655	-33	-35	85	-1068	-1	-1708
	US	-3433	-12	-23	73	-1103	-2	-4499
	G-10	-257	-25	-33	88	-1073	-1	-1302
	G-20	-1547	-25	-31	79	-1082	-2	-2606
Milk	EU	-657	3220	-817	153	-10476	345	-8232
	US	-2158	6979	-1619	164	-11729	502	-7862
	G-10	-366	2409	-672	149	-10162	312	-8330
	G-20	-1096	4014	-926	154	-10797	387	-8264
Beef	EU	-2094	-112	3	-1288	-4585	13	-8062
	US	-8408	-173	4	-1125	-4772	12	-14464
	G-10	-957	-83	-78	-1334	-4544	13	-6982
	G-20	-4391	-135	33	-1219	-4620	13	-10318
Other meat	EU	-38	4802	-3671	231	-3505	14	-2166
	US	3	14636	-7974	235	-3929	14	2986
	G-10	-16	2648	-2861	233	-3404	15	-3385
	G-20	-58	7709	-5237	228	-3622	14	-966

a) For the composition of sectors see Table A4.2 in the appendix.

b) TC = third countries

Source: Authors' calculations.

Table 4.4 also decomposes the change in the EU trade balance for the other meat (pork and poultry) sector, which reveals a large degree of variation between the simulations. The implementation of the G-10 proposal results in a deterioration of the EU trade balance of $-3.39 bn in total, while the EU achieves a relative increase in EU other meat exports of $2.99 bn with the implementation of the US proposal. The decomposition of these results shows that this variation is related to the sensitivity of the EU's other meat sector towards higher agricultural tariff cuts in third countries. The higher agricultural tariff cuts of the US proposal open up the market in third countries, which have to accommodate additional EU exports of $14.64 bn. Yet, only EU exports of $-7.97 bn are displaced by the high cuts to agricultural tariffs in the US proposal and its resulting increased trade opportunities among third countries. An additional relative decrease of EU imports of other meat stems from the abolition of EU export subsidies ($-3.93 bn), but does not compensate this positive effect. On the contrary, the application of the G-10 proposal only implies an increase of exports to third countries of $2.65 bn, while it is just offset by the increase of trade among third countries ($-2.86 bn). The additional negative effect of the elimination of EU export subsidies for other meat rises to $-3.40 bn and leads to an overall deterioration of the EU trade balance for this sector. All other effects are negligible for the EU trade balance of other meat.

The EU sugar sector is another area in which a negative change in the trade balance develops, which lies between $-1.30 bn (G-10 proposal) and $-4.50 bn (US proposal). Analogous to the beef sector, the EU sugar sector reacts very sensitively if the higher agricultural tariff cuts are applied under the US proposal. The EU's relative sugar imports only decrease by a moderate amount $-0.26 bn under the application of the market access options included in the G-10 proposal, while the proposed US tariff cuts lead to a negative change in the EU sugar trade balance of $-3.43 bn. A clearly negative effect also results from the abolition of the EU export subsidy, which, however, is more or less constant among the four simulations and amounts to $-1.07 bn (EU proposal) or $-1.10 bn (US proposal). All proposals result in better access to third countries' sugar markets, but the relative increase in EU sugar exports is only marginal and does not compensate for the other negative effects.

Finally, Table 4.4 also reveals the negative change in the EU trade balance for cereals, rising to $-0.04 bn and $-0.39 bn in the US and the EU proposals respectively. The decomposition of these results shows that there are two reasons for this negative development. First, the EU trade balance for cereals declines owing to the opening of markets for agricultural trade among third countries and its resulting diverting effect for EU cereal exports. This effect involves a negative development in the EU trade balance for cereals of $-0.07 bn under the G-10 proposal, which even increases under the implementation of the US proposal to $-0.51 bn. Second, the EU trade balance for cereals is also hurt by the abolition of EU export subsidies. Here, the trade balance deteriorates by around $-0.8 bn in all four simulations. These negative developments are partly offset by positive developments that result from better EU export opportunities to the markets of third countries by $1.14 bn under the application of the US proposal. Nevertheless, this positive figure falls slightly to $0.46 bn when the agricultural tariff cuts of the G-10 proposal are employed. Table 4.4 shows another positive influence on the EU trade balance for cereals, which stems from the cut of EU agricultural tariffs and lies between $0.09 bn (G-10 proposal) and $0.17 bn (G-20 proposal).

The latter result in particular is somewhat puzzling, and certainly deserves a closer look. Why do the cuts to EU agricultural tariffs have a positive impact on the EU trade balance for cereals? Table 4.5 presents an extended decomposition of this outcome under the US and the G-10 proposals to answer this question.

Table 4.5 Decomposed effects of cuts to EU agricultural import tariffs on the trade balance of cereals (mn US$)

| | Impact of the trade balance of cereals | |
	US proposal	G-10 proposal
Cut of EU tariffs of		
cereals	-303	-10
oilseeds	1	1
paddy rice	12	4
vegetables & fruits	58	18
cattle	0	0
other animal	5	4
beef	149	25
other meat	38	6
milk	33	6
sugar	30	4
other food	78	29
Sum	101	87

Source: Authors' calculations.

As the table shows, the total effect of the cuts to all agricultural tariffs is further broken down into each agricultural commodity. From a first glance it can clearly be seen that the cut to EU import tariffs for cereals undoubtedly has the expected negative effect on the EU trade balance for cereals, under the proposals of the US ($-0.30 bn) and the G-10 ($-0.01 bn). But the cut for all other agricultural products, particularly for beef ($0.15 bn and $0.03 bn respectively), has a positive effect on the EU trade balance for cereals. The reason for this is a reduction of, for example, high EU tariffs on beef imports, which constitute an implicit tax for the EU cereals sector. In summary, these positive effects outweigh the negative effect of the cut in the relatively low EU tariff for cereals. A very similar outcome can be observed when third countries cut their tariffs for EU imports of sugar and beef (see Table A4.4 in the appendix).

4.4.2   Impact on third countries' trade balances

Who is taking advantage of the expanded access to EU agricultural markets? To elaborate this question we divide the countries and regions of Table 4.3 into developed, developing and least-developed countries (LDCs)[11] as well as the rest of the world, which includes non-WTO member countries. Table 4.6 accordingly represents the change in the trade balance for these regions for all the considered commodities. In the following we concentrate the discussion on the highest and lowest figures resulting from each of the four simulations.

An examination of Table 4.6 shows that the previously noted negative changes to the EU trade balance for beef ($-6.98 bn in the G-10 proposal) are mainly accommodated by developing countries ($3.16 bn), while developed countries and the rest of the world share almost all of the remaining surplus of $1.59 and $2.25 bn respectively. Yet only the developed and developing countries are able to increase their beef exports to the EU if the much higher cuts to agricultural tariffs are applied under the US proposal. It is interesting to note that the trade balance of developed countries increases disproportionately here. By contrast, the trade balances of the LDCs change and those of the rest of the world stay relatively constant in the simulations.

The large negative changes of around $-8 bn to the EU trade balance for milk in all the simulations are also distributed among all other groups of countries. Nevertheless, the developing and developed countries can obviously be identified as the main milk-surplus producers, which are able to increase their relative milk exports by $3.6 bn in each case with the help

---

[11] Developed, developing and least-developed countries are grouped according to the WTO classification.

of the tariff cuts proposed by the G-10. Developed and developing countries show, however, a slightly smaller increase in their trade balances if the higher cuts to agricultural tariffs are implemented under the US proposal (about $2.7 bn). The rest of the world is able to almost double an increase in relative exports. Again, the LDCs are only minor players in the world milk market.

Table 4.6 Changes in trade balances owing to the implementation proposals by the EU, the US, the G-10 and the G-20 for aggregated regions (mn US$) [a)]

	EU27	IC	DC	LDC	ROW	EU27	IC	DC	LDC	ROW
	EU proposal					US proposal				
cereals	-388	3615	-3674	83	-58	-39	4239	-5176	138	140
oilseeds	181	582	-972	24	-150	289	1033	-1642	105	-131
paddy rice	-4624	-4715	9409	-1126	-128	-5797	-3910	9815	-1346	-130
vegetables & fruits	-214	-914	1035	-117	-117	-401	-705	438	-96	-55
cattle	176	-722	683	-18	-37	332	-928	782	-14	-38
other animal	-130	-1240	1606	-46	-137	-81	-949	1228	-48	-117
beef	-8053	1433	4335	72	2246	-14466	2178	9560	86	2255
other meat	-2168	-3047	5053	183	-70	2985	-4101	401	266	-82
milk	-8232	3377	3643	588	1037	-7863	2836	2627	643	1793
sugar	-1708	-223	2926	-955	43	-4499	-1326	7742	-2337	102
other food	-7127	-11005	18409	-608	-888	-7234	-11262	17596	-445	-737
other primary	3397	2643	-6125	-1474	-826	3876	2461	-6530	-1140	-859
manufactures	6542	-4244	-23016	4707	1944	8461	-4662	-23862	5314	637
services	22305	14499	-13400	-1304	-2750	24417	15109	-13025	-1132	-2665
	G-10 proposal					G-20 proposal				
cereals	-308	3225	-3184	130	16	-321	3727	-3900	95	-42
oilseeds	90	-767	581	-49	-167	219	705	-1151	40	-145
paddy rice	-883	-8006	8919	-1043	-132	-4651	-4708	9514	-1206	-130
vegetables & fruits	-260	-438	743	-107	-129	-402	-1112	1336	-104	-113
cattle	145	-584	561	-16	-35	238	-737	640	-15	-37
other animal	-107	-1088	1415	-43	-127	-118	-1194	1542	-43	-136
beef	-6976	1588	3161	69	2249	-10301	1420	6504	78	2245
other meat	-3385	-1572	4891	166	-65	-968	-3738	4368	213	-55
milk	-8330	3632	3626	574	975	-8264	2881	3909	606	1198
sugar	-1302	57	1907	-548	29	-2607	-500	4575	-1549	66
other food	-6614	-9432	16409	-572	-639	-7317	-11260	18655	-533	-894
other primary	2796	2998	-5768	-1605	-843	3595	2798	-6543	-1344	-825
manufactures	4319	-3699	-20884	4433	1809	8003	-2921	-25865	5006	1716
services	20763	14130	-12476	-1378	-2838	22854	14669	-13653	-1238	-2740

[a)] For the composition of regions and sectors see Tables A4.1 and A4.2 in the appendix.

Note: IC = industrial countries; DC = developing countries; LDC = least-developed countries; ROW = rest of the world

Source: Authors' calculations.

With regard to the other meat sector, Table 4.6 also reveals that large changes in the trade balances of the EU ($-3.34 bn in the G-10 proposal) and other developed countries ($-1.57 bn) are almost entirely offset by developed countries, which experience a significant positive development in their trade balances ($4.89 bn). Conversely, it is interesting to note that the higher tariff cuts of the US proposal almost exclusively benefit the EU ($2.99 bn), while the other developed countries have to accept a further deterioration of their trade balances for the other meat sector ($-4.1 bn).

The reaction of the sugar sector to the implementation of the Doha round is somewhat different. In this context, the relative increase of EU sugar imports ($-1.30 bn under the G-10 proposal) is accompanied by a trade balance change in the LDCs ($-0.55 bn), which clearly comes from preference erosion. The negative changes to the trade balances of the EU and the LDCs more than double under the application of the US proposal, while the slightly positive change to the sugar trade balances of the developed countries transforms into a relative increase in imports under the US proposal. The main increase of relative exports in the world sugar market is given, however, to developing countries and to a far lesser extent to the rest of the world. Thus, the trade balance change for developing countries ($1.9 bn) under the G-10 proposal is more than doubled when tariff cuts are more ambitiously applied with the implementation of the US proposal ($7.7 bn). The rest of the world is only a casual bystander in the world sugar market.

Finally, Table 4.6 reveals the relative increase of cereal imports into the EU as previously indicated. In this respect, the application of the EU proposal leads to a change in the trade balance for the EU and the developed countries of $-0.39 bn and $-3.67 bn, respectively. These changes are almost completely accommodated by developed countries ($3.62 bn). Owing to better access to developing countries' markets, the change to the EU trade balance is slightly less when the higher cuts to agricultural tariffs are applied as proposed by the US.

Conclusions

The WTO negotiations of the Doha round are a key issue in the public debate. This chapter analyses the effects of different market access options on the basis of a general equilibrium model. An extended version of the GTAP model is used to first project a base run that includes Agenda 2000, the EU's enlargement, the EBA agreement and the MTR. The policy simulation run also includes the WTO negotiations. For this latter aspect, a distinction is made among the four different simulations examining the approaches to open agricultural markets as proposed by the EU, the US, the G-20 and the G-10 in advance of the WTO's Ministerial Conference in Hong Kong in October 2005. All tariff cuts are calculated in the tariff database (at the HS 6-digit tariff level), taking applied and bound rates into account and added up to the GTAP-model level using import weights. In our examination, we concentrate on the impact of the WTO negotiations on the EU's trade flows. In this respect, the results of the simulations reveal the following points:

- Results from different options for market access in the WTO negotiations of the Doha round show parallel developments. Thus, for example, the increase or decrease of the trade balance is more or less pronounced, while a change of direction is merely an exception that applies to the other meat sector.

- Implementation of all the proposals results in negative changes to the EU's trade balances for most agricultural products, but particularly for the highly protected EU beef and milk sectors. The trade balances of sugar, cereals, vegetables and fruit, other food products and to some extent those for other meat products deteriorate as well, and consequently show a relative increase in imports. These developments are only reversed for other meat products, if higher agricultural tariffs (e.g. as proposed by the US) are implemented.

- With the help of a decomposition of the trade effects for selected products it can be shown that the negative developments in the EU trade balances for beef and sugar are dominated by the cuts in EU agricultural tariffs and the abolition of EU export subsidies. For the milk trade balance, however, the largest negative trade impact for the EU stems solely from the abolition of EU export subsidies, while tariff cuts by the EU and third countries are not so important. The EU cereals and other meat sectors particularly react to tariff cuts among third countries, which displace EU exports to them. The elimination of export subsidies is also of significant importance for the EU cereals and other meat sectors.

- Who is taking advantage of improved EU market access? From the perspectives of the non-participating LDCs and non-WTO member countries, it does not make much difference whether tariff cuts are higher (as in the US proposal) or not (as in the G-10 proposal). These proposals only realise minor changes in trade balances. Additionally, the LDCs suffer from preference erosion in the sugar sector, which increases with higher tariff cuts. By contrast, developing countries are able to disproportionately increase their relative beef and sugar exports to the EU. This development even occurs when the tariff cuts are higher. Other developed countries are able to increase their relative cereal exports, but only gain slightly better access to other EU agricultural markets.

- From the EU's point of view, it really matters which of the four proposals are implemented in the beef, other meat and sugar sectors, as these sectors are particularly sensitive to tariff cuts by the EU and third countries. For the EU milk sector, however, it hardly makes a difference, as better market access to third countries is almost entirely offset by the effect of tariff cuts by the EU and among third countries.

References

Anderson, K. and W. Martin (2005), "Scenarios for Global Trade Reform", in W. Hertel and A. Winters (eds), *Putting Development Back into the Doha Agenda: Poverty Impacts of a WTO Agreement*, World Bank, Washington, D.C.

Bouët, A., Y. Decreux, L. Fontangné, S. Jean and D. Laborde (2004), *A Consistent Ad-Valorem Equivalent Measure of Applied Protection across the World, the MAcMap-HS6 database*, CEPII Working Paper No. 2004-22, CEPII, Paris.

Brockmeier, M. (2003), *Ökonomische Auswirkungen der EU-Osterweiterung auf den Agrar- und Ernährungssektor - Simulationen auf der Basis eines Allgemeinen Gleichgewichtsmodells*, Habilitation, Wissenschaftsverlag Kiel, Agrarökonomische Studien, No. 22, Kiel.

Brockmeier, M., U. Sommer and K. Thomsen (2005), "Sugar policies: An invincible bastion for modellers?", in F. Arfini (ed.), *Modelling Agricultural Policies: State of the Art and New Challenges, Proceedings of the 89th European Seminar of the European Association of Agricultural Economists (EAAE)*, Parma, 3–5 February 2005.

FAPRI (2005), *U.S. Proposal for WTO Agricultural Negotiations: Its Impact on U.S. and World Agriculture*, CARD Working Paper No. 05-WP 417, FAPRI, Ames, Iowa.

Frandsen, F., B. Gersfeld and H. Jensen (2002), *Decoupling Support in Agriculture: Impacts of Redesigning European Agricultural Support*, paper presented at the 5th Annual Conference on "Global Economic Analysis", Taipei, 5–7 June (retrieved from www.gtap.agecon. purdue.edu/resources).

Francois, J.F. and W. Martin (2003), "Formulas for Success? Formula Approaches to Market Access Negotiations", *World Economy*, Vol. 26, No. 1, pp. 1-28.

G-10 (2005), *G-10 Proposal on Market Access*, 10 October (retrievable from http://www.ictsd.org/ministerial/hongkong/documents_resources. htm).

G-20 (2005), *G-20 Proposal on Market Access*, 12 October (retrievable from http://www.ictsd.org/ministerial/hongkong/documents_resources. htm).

Harrison, W.J., J.M. Horridge and K.R. Pearson (1999), *Decomposing Simulation Results with Respect to Exogenous Shocks*, Working Paper No. IP-73, CoPS/IMPACT, Clayton, Victoria, Australia.

Hertel, T.W. (ed.) (1997), *Global Trade Analysis: Modelling and Application*, Cambridge, MA: Cambridge University Press.

Martin, W. and Z. Wang (2004), "Improving Market Access in Agriculture", mimeo, World Bank, Washington, D.C.

Podbury, T. and I. Roberts (2003), *Opening Agricultural Markets through Tariff Cuts in the WTO*, ABARE Report No. 03.2, RIRDC publication No. 03/011, Canberra.

USTR (2005), *US Proposal for WTO Agriculture Negotiations*, Office of the US Trade Representative, 10 October.

Walkenhorst, P. and N. Dihel (2003), *Tariff bindings, unused protection and agricultural trade liberalization*, OECD Economic Studies No. 36, 2003/1, OECD, Paris.

Walmsley, T.L., B.V. Dimaranan and R.A. McDougall (2000), *A Base Case Scenario for the Dynamic GTAP Model*, Centre for Global Trade Analysis, Purdue University, West Lafayette, IN.

WTO (2001), Ministerial Declaration, WT/Min(01)/DEC/1, WTO, Geneva (retrieved from http://www.wto.org).

————— (2003a), Negotiations on Agriculture: First Draft of Modalities for the further Commitments, Revision TN/AG/W/1/Rev1, WTO, Geneva (retrieved from http://www.wto.org).

————— (2003b), EU–US Joint Text, Agriculture, JOB(03)/157, WTO, Geneva (retrieved from http://www.wto.org).

————— (2003c), Draft Cancún Ministerial Text, WTO, Geneva, 24 August (retrieved from http://www.wto.org/english/thewto_e/ minist_e/ min03_e/draft_decl_e.htm).

————— (2003d), Preparations for the Fifth Session of the Ministerial Conference, Draft Cancún Ministerial Text, Second Revision, JOB(03)/150/Rev2, WTO, Geneva (retrieved from http://www. wto.org).

————— (2004a), Doha Work Programme, Draft General Council Decision of 16 July 2004, JOB(04)/96, WTO, Geneva (retrieved from http:// www.wto.org/english/tratop_e/dda_e/ddadraft_16jul04_e.pdf).

————— (2004b), Doha Work Programme, Draft General Council Decision of 30 July 2004, Revision JOB(04)/96/Rev1, WTO, Geneva, (retrieved from http://www.wto.org).

————— (2004c), Doha Work Programme, Decision adopted by the General Council on 1 August 2004, WT/L/579, WTO, Geneva (retrieved from http://www.ige.ch/E/jurinfo/documents/j10407e .pdf).

Appendix

Table A4.1 Aggregation of countries and regions

Countries and regions	Abbr.
1. European Union – 27 countries Austria, Belgium, Bulgaria, the Czech Republic, Cyprus, Denmark, Estonia, Finland, France, Germany, Greece, Hungary, Ireland, Italy, Latvia, Lithuania, Luxembourg, Malta, the Netherlands, Poland, Portugal, Romania, Slovakia, Slovenia, Spain, Sweden, the United Kingdom	EU-27
2. United States	USA
3. Japan	JPN
4. Oceania Australia, New Zealand	OCEA
5. Other WTO members (industrialised countries) Canada, Switzerland, Rest of EFTA, Albania, Croatia	WTO IC
6. Brazil	BRA
7. India	IND
8. Rest of African, Caribbean and Pacific countries Rest of Oceania, Rest of FTAA, Rest of Caribbean, Botswana, South Africa	ACPs
9. Least-developed countries Bangladesh, Rest of Southeast Asia, Rest of South Asia, Malawi, Mozambique, Tanzania, Other Southern Africa, Madagascar, Uganda, Rest of Sub-Saharan Africa, Zambia	LDCs
10. Other WTO members (developed countries) China, Hong Kong, Korea, Rest of East Asia, Indonesia, Malaysia, Philippines, Singapore, Thailand, Sri Lanka, Mexico, Colombia, Peru, Venezuela, Rest of Andean Pact, Argentina, Chile, Uruguay, Rest of South America, Central America, Turkey, Rest of Middle East, Morocco, Tunisia, Rest of North Africa, Rest of South African CU, Zimbabwe	WTO DC
11. Rest of the world Taiwan, Vietnam, Rest of North America, Rest of Europe, Russian Federation, Rest of FSU	ROW

Source: Authors' compilation.

Table A4.2 Aggregation of sectors

Sectors

1. Wheat, cereal grain nec

2. Oil seeds

3. Sugar cane, sugar beet

4. Paddy rice

5. Vegetables, fruit, nuts

6. Cattle, sheep, goats, horses

7. Animal products nec

8. Raw milk

9. Meat: cattle, sheep, goats, horses

10. Meat products nec

11. Dairy products

12. Sugar

13. Food products nec, vegetable oils and fats, processed rice

14. Other primary sectors

Plant-based fibres, crops nec, wool, silk-worms, cocoons, forestry, fishing coal, oil, gas, minerals nec, wood products, petroleum, coal products

15. Industry

Beverages and tobacco products, textiles, wearing apparel, leather products, wood products, paper products, publishing, chemical, rubber, plastic products, mineral products nec, ferrous metals, metals nec, metal products, motor vehicles and parts, transport equipment, electronic equipment, machinery and equipment, manufactures nec

16. Services

Electricity, gas manufactures, distribution, water, construction, trade, transport nec, sea transport, air transport, communication, financial services nec, insurance, business services nec, recreation and other services, public administration/defence/health/education, dwellings

Source: Authors' compilation.

Table A4.3 Pre-simulations, Agenda 2000 and EU enlargement

Pre-simulations

CAP instruments

- Complementarity approach for milk and sugar (assumption: quantity in the database represents production quotas)
- Land subsidy equalised across sectors to implement a homogeneous area payment

Common EU budget

- 75% of tariff revenues as well as a share of GDP is accrued to the EU budget; determination of a uniform endogenous GDP rate
- Expenses of the EAGGF paid for by the common EU budget
- Net transfers among EU member states

Agenda 2000

Cereals

- Reduction of intervention prices by –15%
- Unification of direct payments for cereals, oilseeds and protein plants
- Reduction of set-aside rate from 15% to 10%

Beef

- Reduction of intervention prices by –18%
- No change in direct payments (assumption: increase in direct payments is compensated by lower output)

Milk

- Reduction of intervention prices by –15%
- Retention of quota regulation
- Increase of quota by 2.4%

EU enlargement (EU-27)

Creation of customs union

- EU-15 and CEECs abolish all bilateral trade barriers
- The CEECs establish the trade protection measures of the EU-15
- Production quotas for milk and sugar are fixed at the current production levels of the CEECs
- No set-aside in the new member states

Table A4.3 Continued

- ▪ Direct payments in the EU-15 remain unchanged
- ▪ 100% of the current land and animal premiums in the EU-15 are transferred to the new member states (standard procedure)
- ▪ Fixation of ceilings for direct payments with an endogenous adjustment of the premium rate for land and animals in the EU-15

Common EU budget

- ▪ Complete integration of the CEECs in the EU's common budget: 90% of tariff revenues as well as a share of GDP to the EU budget
- ▪ Payments in the framework of the EAGGF in the CEECs through the common budget
- ▪ Implementation of net transfers between the EU-15 and the CEECs

Source: Authors' compilation.

Table A4.4 Decomposed effects of third countries' tariff cuts for EU agricultural imports on the trade balance of sugar and beef (mn US$)

	Impact of the trade balance of *sugar*		Impact of the trade balance of *beef*	
	US proposal	G-10 proposal	US proposal	G-10 proposal
Cut of third countries tariffs for EU imports of				
cereals	-8.2	-2.0	-111.0	-39.0
oilseeds	-0.4	0.0	-7.0	-2.0
paddy rice	0.0	0.0	0.0	0.0
vegetables & fruits	-9.2	-2.0	-95.0	-19.0
cattle	-0.4	0.0	-11.0	-5.0
other animal	-1.9	-0.7	-8.0	-3.0
beef	-2.0	0.0	560.0	100.0
other meat	-53.7	-10.0	-316.0	-54.0
milk	-37.3	-11.0	-118.0	-40.0
sugar	145.0	14.0	-1.0	0.0
other food	-44.2	-13.0	-66.0	-21.0
Sum	-12.3	-24.7	-173.0	-83.0

Note: For the composition of the sectors see Table A4.2 above.

Source: Authors' calculations.

Table A4.5 Changes in output owing to the implementation of proposals for market access by the EU, the US, the G-20 and the G-10 (in %)

	EU27	USA	Japan	Oceania	WTO IC	Brazil	India	ACP	LDC	WTO DC	ROW
EU proposal											
cereals	-4	6	-39	-1	2	15	0	0	1	-9	1
oilseeds	0	4	5	-17	-21	18	-1	-5	0	-24	-4
sugar beet & cane	-14	-1	-18	4	2	0	0	17	-3	6	1
paddy rice	-35	-6	-87	7	-28	-15	-1	-7	-8	-46	-2
vegetables and fruits	-1	-4	-1	-1	4	-6	-1	-2	0	3	0
cattle	-13	0	-15	-1	4	14	0	1	0	5	3
other animal	-2	-2	-12	-6	-11	-9	0	-1	2	5	0
raw milk	-6	1	-3	17	4	1	1	6	4	7	-1
beef	-28	2	-18	8	1	28	-3	1	3	5	1
other meat	-3	1	-42	-4	-20	-23	18	-2	6	12	44
milk	-10	1	-3	20	7	1	6	14	75	13	4
sugar	-21	-1	-18	14	2	1	0	29	-8	8	16
other food	-3	-2	-4	1	-2	-4	-6	-1	-1	10	2
other primary	1	0	0	0	0	-1	-1	0	-1	0	-1
manufactures	0	0	1	-2	-1	-1	1	-1	2	-1	0
services	0	0	0	0	0	0	0	0	0	0	0
US proposal											
cereals	-3	7	-75	-1	3	18	0	-3	1	-12	1
oilseeds	2	5	8	-22	-21	10	-1	-7	2	-23	-3
sugar beet & cane	-28	-4	-36	12	-5	1	0	58	-7	9	1
paddy rice	-46	-1	-87	9	-35	-18	0	2	-9	-40	-2
vegetables and fruits	-2	-4	-1	-2	9	-12	-3	-2	0	3	0
cattle	-21	1	-30	6	6	50	-6	0	0	5	3
other animal	0	0	-23	-7	-11	-9	0	-1	2	5	2
raw milk	-7	1	-9	35	-15	-1	1	3	4	6	2
beef	-41	4	-36	15	-1	87	-22	0	3	5	44
other meat	2	4	-76	-8	-22	-38	-30	-9	8	8	4
milk	-10	2	-11	41	-22	-1	5	8	79	12	25
sugar	-40	-5	-37	41	2	6	0	104	-20	14	3
other food	-3	-2	-4	3	-1	-4	-8	-3	-1	10	-1
other primary	1	0	0	-2	0	-5	0	0	0	0	0
manufactures	0	-1	2	-4	-1	-4	2	-1	2	-1	-1
services	0	0	0	0	0	0	0	0	0	0	0

1) For the composition of regions and sectors see Table A1 and A2 in the appendix. 2) WTO IC: other developing WTO member countries, ACP: African, Caribbean and Pacific countries; LDC: Least Developed countries; WTO DC: other developing WTO member countri

Table A4.5 Continued

	EU27	USA	Japan	Oceania	WTO IC	Brazil	India	ACP	LDC	WTO DC	ROW
G-10 proposal											
cereals	-4	5	3	2	5	-4	0	3	1	-1	1
oilseeds	-1	-1	2	-16	-22	23	0	-5	-1	-20	-4
sugar beet & cane	-11	0	-11	1	1	0	0	-8	-1	5	1
paddy rice	-7	-17	-82	-13	-8	-13	-2	-15	-7	-46	-2
vegetables and fruits	-1	-2	0	-1	0	-4	0	-2	0	2	0
cattle	-10	1	-12	-1	3	5	0	1	1	4	3
other animal	-3	-1	-9	-5	-4	-8	1	-1	1	5	2
raw milk	-6	1	-1	15	6	1	1	6	4	6	1
beef	-24	2	-14	6	3	13	2	1	3	4	44
other meat	-4	2	-32	-2	-5	-17	87	-1	6	10	4
milk	-10	1	0	18	11	1	6	15	74	13	16
sugar	-16	0	-11	8	1		-4	15	-4	7	1
other food	-3	-2	-3	-2	-2	-3	-2	-1	-1	9	-1
other primary	1	0	0	0	0	0	1	0	-1	0	-1
manufactures	0	0	1	-2	-1	0	0	-1	2	0	0
services	0	0	0	0	0	0	0	0	0	0	0
G-20 proposal											
cereals	-4	6	-46	-1	2	12	0	0	1	-8	1
oilseeds	1	4	6	-18	-21	15	-1	-5	1	-24	-3
sugar beet & cane	-20	-2	-22	6	2	0	0	31	-5	7	1
paddy rice	-35	-6	-87	6	-27	-16	-1	-5	-8	-46	-2
vegetables and fruits	-2	-4	-1	-2	4	-8	-3	-2	0	3	0
cattle	-16	1	-21	-1	5	29	0	1	0	5	3
other animal	-2	-1	-17	-6	-11	-8	-1	-1	2	5	3
raw milk	-7	1	-5	22	-2	0	1	6	4	7	1
beef	-33	3	-25	8	1	52	-12	1	3	6	44
other meat	-2	3	-65	-4	-21	-28	-5	-5	7	12	4
milk	-10	1	-6	26	-2	0	6	14	76	14	18
sugar	-29	-2	-22	19	1	3	0	56	-13	10	2
other food	-3	-2	-4	2	-2	-4	-5	0	-1	10	-1
other primary	1	0	0	-1	0	-2	-1	-1	-1	0	-1
manufactures	0	0	1	-3	-1	-2	0	0	2	-1	0
services	0	0	0	0	0	0	0	0	0	0	0

1) For the composition of regions and sectors see Table A1 and A2 in the appendix. 2) WTO IC: other developing WTO member countries, ACP: African, Caribbean and Pacific countries; LDC: Least Developed countries; WTO DC: other developing WTO member countri

Source: Authors' calculations.

5. An economy-wide perspective on Euro-Mediterranean trade agreements with a focus on Morocco and Tunisia

Marijke Kuiper and Frank van Tongeren[1]

5.1 The issues at stake

The economic interests of the EU and the Mediterranean partner countries (MPCs) in the Euro-Mediterranean Association Agreements (EuroMed Agreements) are far apart. The MPCs are of limited economic interest to the EU. Imports from the MPCs account for only 2% of EU imports, while exports to the MPCs account for only 3% of total EU exports. The majority of EU imports from the MPCs consist of oil, followed at a distance by Mediterranean agricultural products. Given the limited size of the agrarian trade flows, the European Commission does not consider the MPCs a threat to European farmers (Garcia-Alvarez-Coque, 2002).

The very limited economic interests of the EU contrast with the clear economic interests of the MPCs: 50% of their imports and exports are with the EU, which is their largest trade partner. The MPCs have a comparative advantage in typical Mediterranean products such as fresh fruit and vegetables, citrus, tomatoes and olive oil. Improved access to European

[1] This chapter benefited greatly from detailed comments by Mohamed Lahouel during the ENARPRI final conference "Trade Agreements and EU Agriculture" held in Brussels on 8 June 2006. Financial support through the EU-sponsored ENARPRI-TRADE project is gratefully acknowledged, as is financial support from the Dutch Ministry of Agriculture, Nature Management and Food Quality.

agricultural markets could provide a positive stimulus to their economies. Such a positive stimulus is badly needed. Economic growth in the MPCs is lagging behind the growth rates attained in the rest of the world, while the MPCs combine a young population with unemployment rates of between 15 and 30%.

Given the limited economic interests in the MPCs, the EuroMed Agreements de facto mainly serve the EU's political interests in stability at its southern borders. This political interest can also be inferred from explicit references to the Barcelona process in the European security strategy launched by High Representative Javier Solana (European Council, 2003). A further indication of the EU's political interests in the Mediterranean is a speech on the link between the European Neighbourhood Policy and the Euro-Mediterranean Partnership (Wallström, 2005). Looking at the EuroMed Agreements in terms of promoting political interest in stability and prosperity at the EU's southern borders, there appears a contradiction between the political interests of the EU as a whole and the sectoral or regional economic interests of specific EU member states.

The EU unilaterally removed its protection on manufactured goods in the 1970s, but has since maintained its high levels of protection in agriculture. The EuroMed Agreements thus boil down to an opening of MPC markets to industrial imports from the EU. Since the MPCs' industrial producers are not generally considered internationally competitive, implementation of the agreements is expected to decimate industry in the MPCs. The resulting reduction in already limited employment will not contribute to stability in the MPCs.

Next to an expected loss of employment in the industrial sector, implementation of the agreements will decrease tariff revenues. Government expenditures in the MPCs are high owing to a bloated public sector: the share of non-military public employment in total employment is twice the world average (Bulmer, 2000). In addition, producers of grain, meat and milk are subsidised to reduce dependence on imports, while consumer prices are kept low through subsidies on staple foods. A reduction of government revenues through trade liberalisation would have a direct impact on employment and consumer prices, with all its consequences in terms of social unrest.

The current agreements thus conflict with the political interests of the EU in attaining stability at its southern borders, by having detailed schemes for abolishing protection on manufactured goods, but so far not on agricultural products. From the perspective of stability at the southern perimeters, and given the comparative advantage of the MPCs in Mediterranean agriculture, the trade agreements should aim at relaxing the complex EU trade barriers for Mediterranean agricultural products. Current EU concessions in this area are marginal, since MPC producers compete directly with producers from Mediterranean EU member states. Although concessions would have only a marginal impact on the EU as a whole, relaxing restrictions on Mediterranean agricultural products would be noticeable in Mediterranean EU member states. The current trade agreements reflect these regional interests.

European trade barriers, however, are only one of the factors limiting economic growth in the MPCs. Next to a bloated public sector and market interventions, the region belongs to the most protected in the world and the competitiveness of the private sector is limited by the high level of trade protection. In addition, there is an inflow of foreign exchange in the MPCs through oil revenues and remittances. This inflow of foreign exchange stimulates domestic demand for services and causes an appreciation of the exchange rate by increasing demand for imports, thus hampering exports. As a result, MPC economies are oriented towards non-trade sectors.

Trade protection distorts the structure of the economy while creating interests in maintaining the protection that hampers reform. An example is the industrial sector in the MPCs. Access to European markets has not led to a competitive sector, because continuing high trade barriers have not provided an incentive to restructure industries. A comparable scenario would be possible with unconditional and unilateral reduction of European trade barriers for agricultural products. The complexity of the current protection implies that producers have invested in information and contacts to be able to export to the EU and thus they have an interest in maintaining the current protection structure. If the MPCs retain their barriers to imports from the EU, there are no incentives for restructuring agricultural production.

Given the limited economic interest of the MPCs for the EU as a whole, there is room to support domestic reforms in the MPCs through alignment of trade agreements with domestic policies. Consequently, the aim of this chapter is to 1) analyse the impact of the EuroMed Agreements on northern and Mediterranean EU member states and 2) assess the interactions between the agreements and domestic policies in the MPCs.

This study is structured as follows. Section 5.2 discusses the model and data used in the study. We use a computable general equilibrium (CGE) model that allows us to analyse changes in trade flows and the resulting economy-wide effects in Morocco and Tunisia (as case studies of the MPCs) and in northern and Mediterranean EU member states. Section 5.3 describes the scenarios analysed in this study, as well as initial tariff profiles and trade flows. To gauge the impact of assumptions on model outcomes we focus on extreme assumptions, providing upper and lower bounds of possible outcomes. In terms of the scope of liberalisation we analyse the current EuroMed Agreements (limited to an opening of MPC economies to EU manufacturers) and a full liberalisation of all trade (including agriculture). In terms of domestic policies we analyse a case in which tariff revenues are not replaced (government expenditures are assumed to adjust) and a case with full replacement of tariff revenues by a consumption tax. Together this yields four scenarios, varying in terms of the scope of liberalisation and tariff replacement. Section 5.4 analyses the impact of the current EuroMed Agreements, with and without tariff replacement. Section 5.5 assesses whether the inclusion of agriculture in the agreements would lead to different results. The final section concludes.

5.2   Model and data

The issue at stake in this study is to analyse the economy-wide effects of the EuroMed Agreements, taking the diverging interests of EU member states into account. This effort poses a set of requirements on the applied methodology. Kuiper (2004) provides a review of existing general equilibrium analyses of the EuroMed Agreements in light of these requirements. Here we focus on the main implications of this review for the modelling exercise in this study.

In order to address the diverging economic interests of EU member states a multi-regional model is required. Similarly, in order to address the prospects of South–South integration to balance the hub-and-spoke nature of the bilateral agreements between the EU and the MPCs such a model is needed.

In this study we employ a multi-regional general equilibrium model with the Global Trade Analysis Project (GTAP) database (version 6). While being somewhat restricted in terms of sectoral detail, an economy-wide analysis does allow an examination of the trade-offs between agricultural and non-agricultural sectors and the employment effects of the agreements.

The current EuroMed Agreements are most detailed in terms of lowering restrictions on trade in manufactured goods by the MPCs, and we therefore feel that a more aggregated but economy-wide perspective on the EuroMed Agreements in the context of a multi-regional model is warranted.

5.2.1   Main model features

As noted above, to gauge the economy-wide effects of the EuroMed Agreements we use a CGE model. The GTAP model we employ for this purpose provides a complete representation of each national economy as well as their interactions through trade. In terms of production it covers the agricultural, manufacturing and services sectors, allowing an analysis of resource reallocation following policy changes. In terms of incomes and consumption it covers the use of factor income, tariff revenues and taxes for private and government expenditures. The key strength of the GTAP model is the detailed modelling of international trade, including all bilateral trade flows among regions. By distinguishing bilateral trade flows we can analyse the impact of preferential trade agreements, of which the EuroMed Agreements are an example. For a detailed discussion of the algebraic structure of the GTAP, see chapter 2 in Hertel (1997).

Whereas the GTAP has much to offer in terms of coverage of production and trade, consumption is modelled in less detail. Consumption decisions are modelled using a single household to represent all consumption decisions in a region. Consequently, the results do not provide much in terms of the distribution of income across households, nor can we analyse changes in poverty. In terms of changes in the distribution of income we can analyse the returns to land, labour (skilled and unskilled) and capital, which provide a limited indication of changes in income across different household groups.

To capture the importance of unemployment in the MPCs we use an unemployment closure, replacing the standard assumption of perfect labour markets. The standard assumption implies that full employment is attained by adjusting the wage rate until demand and supply are balanced. In contrast, the unemployment closure specifies that real wages are fixed at base levels and the (perfectly elastic) supply follows demand. This is a rather stylised representation of unemployment, which does not take actual levels of unemployment into account. That being said, it does provide a view of the direction of changes in employment as a consequence of changes in policy, thus allowing us to assess whether employment will increase or decrease in different sectors of the economy.

Tariff revenues are an important source of MPC governments' budgets. The elimination of tariffs implies a reduction in revenues, which either requires an adjustment in government expenditures or a tax increase to replace tariff revenues. We analyse the impact of different choices regarding the loss of tariff revenues by varying assumptions on tariff replacement across scenarios. Specifically, we analyse the impact of a tax replacement scenario whereby the consumption taxes are endogenously adjusted such that total taxes are kept at a constant share of national income.

5.2.2   Data, aggregation and baseline construction

Our model employs version 6 (public release) of the GTAP database, representing the economy of 87 regions in 2001 (Dimaranan & McDougall, 2005). Our aggregation of regions for this study is driven by the diverging interests between northern and Mediterranean EU member states, which are specified as two separate regions in the model. We furthermore distinguish EU accession countries in order to account for the recent accession of the new EU member states as well as the imminent accession of Bulgaria and Romania. MPCs are represented by Morocco, Tunisia and the rest of North Africa aggregate, which comprises Algeria, Egypt and Libya. We also distinguish among the rest of the Middle East, which includes some of the MPCs, Israel, Jordan, Lebanon, the Palestinian Authority and Syria, as well as other countries in the Middle East. This distinction allows us to analyse the impact of the EuroMed Agreements on South–South trade. Finally, to allow future analyses of the impact of trade agreements between the MPCs and the US we keep the US separate from the rest of the world.

The sector aggregation is based on an analysis of the scope of the current EuroMed Agreements and the expected employment repercussions. We combined detailed trade data on EuroMed Agreements at the HS-6 level to factor shares from the GTAP to determine an appropriate grouping of sectors. Vegetables and fruit are kept as a separate sector because of the comparative advantage of the MPCs in these products; spices and other crops are kept separate because of different patterns in the proposed elimination of protection. Tariff reductions in manufacturing do not show much variance across sectors, apart from different types of equipment sectors, which are thus kept as a separate sector. The remaining sectors are grouped according to the share in production to be able to capture differences in employment across sectors. After aggregating regions and sectors we arrive at a model with 9 regions and 17 sectors (see the chapter appendix for the aggregation schemes).

The GTAP data represent the state of the world economy in 2001. Meanwhile, other policy changes affect the results that could be attained by the EuroMed Agreements. To account for these policy changes we construct a baseline, updating the 2001 data with the following policy changes in the period 2001–07:

- *the General Agreement on Tariffs and Trade (GATT) and other World Trade Organisation (WTO) agreements* – implementation of the remaining Uruguay round commitments (with an end date of 2005 for developing countries) and the Agreement on Textiles and Clothing (phasing out restrictions on textile trade from 2005 onwards);

- *China's WTO accession* – the accession of China to the WTO is incorporated by equalling all import tariffs according to the most-favoured nation clause;

- *enlargement of the EU in 2004* – expansion of the EU is simulated as full trade liberalisation between the EU-15 and EU-25 and adoption of EU border tariffs in the Central and Eastern European countries; and

- *mid-term review of the common agricultural policy (CAP)* – the GTAP data do not include the reform of the CAP under the Luxembourg Agreement of 2003. The crucial point of this mid-term review is the decoupling of payments from output levels. This reform is approximated by equalising subsidy payments to land and capital across crops. There is thus no gain in terms of subsidies to switch from one crop to another.

The model results after implementing the above set of reforms are used as the reference point for the simulations in this study. This allows us to assess the impact of the EuroMed Agreements in the context of ongoing policy reforms. This study distinguishes rather aggregate sectors, which ignores some of the complexities of the protection structure for agricultural products. The results should thus be interpreted as an indication of the scale and direction of the anticipated effects of the EuroMed Agreements, aimed at providing input for deciding among different proposals for liberalising trade.

5.3   Tariff structures, model scenarios and initial trade flows

The tariff reductions agreed upon in the current EuroMed Agreements focus on manufactured goods. There are a limited number of concessions for agricultural products, consisting of enlargements of quota and the reduction of some tariffs. Most of these imply changes to the complex protection for specific agricultural products, which we cannot adequately capture with the aggregated representation of these crops in the GTAP database. Our discussion of the current EuroMed Agreements thus focuses on changes in the protection of manufactured goods, ignoring the (limited) changes in the protection of agricultural goods.

The EuroMed Agreements contain detailed schemes for lowering import tariffs on manufactured goods by Morocco and Tunisia over a period of up to 12 years. Schemes for the elimination of trade barriers vary from immediate elimination of tariffs, stepwise elimination over a short period (three to five years), stepwise elimination over a long period (up to 12 years or during part of a 12-year period), elimination of tariffs after 12 years and products that are exempted from tariff reductions. The specifications for reductions to be applied by the EU are much simpler: immediate elimination of all protection on manufactured goods except for the protection on the agricultural component of imported goods. This approach reflects the lowering of trade barriers on manufactured imports from MPCs in the past, implying that a very limited number of barriers for manufactured goods from the MPCs remain.

The impact of the EuroMed Agreements depends on the change in prices of EU imports compared with imports from other regions not benefiting from the tariff reduction. Figures 5.1a and 5.1b therefore present tariffs levied by source of imports for Morocco and Tunisia, which are calculated from trade and tariff data at the HS-6 level of detail. [2]

---

[2] The computation of tariff reductions foreseen in the EuroMed Agreements is done by combining 6-digit level data on tariff lines to be removed from Annexes 3 through 6 in the EuroMed Agreements of Morocco and Tunisia with tariff and trade data from MAcMAP. We first compute the trade-weighted import tariff on imports originating in northern Mediterranean EU countries at the 6-digit level using the MAcMap database (these data are from 2001 and thus cover the period before EU enlargement). We then compute the trade-weighted tariff for each of the sectors in our model. The next step involves removing tariffs in accordance with

For the EU we include the tariffs before the EuroMed Agreements as well as after them, in order to assess changes to the relative trade barriers.

Morocco and Tunisia both have significant protection levels across the board. The two countries also tend to levy relatively high tariffs on imports from other trade partners in the region. For agricultural sectors, when comparing the pattern of protection we find generally stronger protection in Tunisia, while Morocco levies higher tariffs on non-agricultural sectors. Given that the EuroMed Agreements mostly affect the non-agricultural sectors we would thus expect a stronger impact of the current agreements on Morocco. The inclusion of agriculture in the full liberalisation scenario, on the other hand, would affect Tunisia more strongly.

Analysing the reduction in tariff barriers on European imports in relation to tariffs levied on imports from other regions, we predictably find only minor changes in the agricultural sectors. In the non-agricultural sectors, however, there is a change in the relative barriers of the EU compared with imports originating in Tunisia or in the rest of North Africa. Whereas the EU encounters higher barriers before the EuroMed Agreements, after the agreements it faces no tariffs (except for a small remaining import tariff on European equipment in Morocco and on European motor vehicles in Tunisia). The relative competitiveness of EU imports compared with imports from other North African countries improves, reflecting the hub-and-spoke structure of the EuroMed Agreements. This structure may redirect trade flows to the EU, weakening economic integration within North Africa.

---

the EuroMed Agreements at the 6-digit level and re-computing the average trade-weighted tariff for imports originating in the EU for the sectors in our model. Afterwards, we compute the percentage change in tariffs for each of the sectors in our model, which renders the shocks to be applied to simulate the EuroMed Agreements.

Figure 5.1a Import tariffs by source of imports levied by Morocco (%)

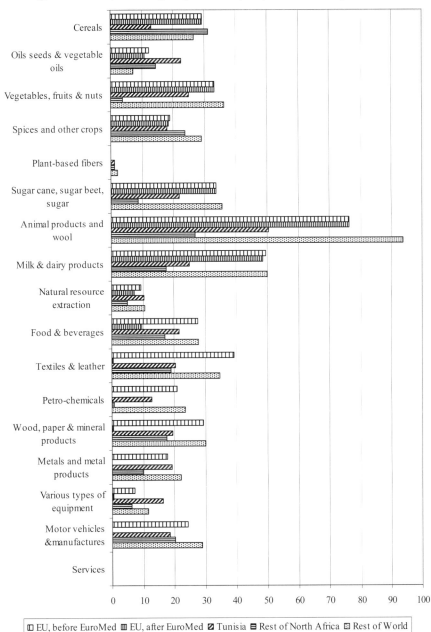

Note: Tariffs are weighted by import flows in the baseline.

Sources: GTAP data and authors' calculations.

Figure 5.1b Import tariffs by source of imports levied by Tunisia (%)

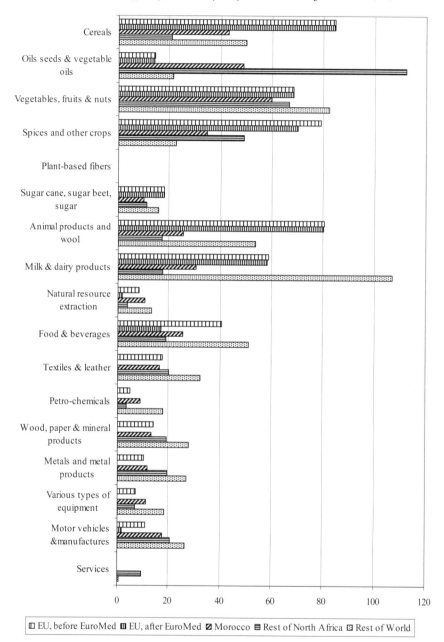

Note: Tariffs are weighted by import flows in the baseline.

Sources: GTAP data and authors' calculations.

These findings for the manufacturing sectors are not unexpected, given the structure of the agreements. More notable is the outcome for the food processing sector – after the EuroMed Agreements the EU also benefits from lower tariffs than those levied on other North African countries, with its tariffs reduced by more than 50% in both Morocco and Tunisia.

Figures 5.1a and 5.1b are based on our computation of the changes in tariffs as foreseen by the EuroMed Agreements. Apart from a clause on maintaining the protection on the agricultural components of manufactured goods, the EuroMed Agreements stipulate an immediate elimination of EU barriers on imports from Morocco and Tunisia. Since we lack information on the agricultural component of manufactured goods we implement this requirement as a straightforward elimination of the remaining (low) barriers on the EU's imports of manufactured goods from Morocco and Tunisia. By ignoring the agricultural component there is the implication that we will slightly overstate the effect of EU liberalisation, but given the low initial tariffs we do not expect this to significantly affect the results.

Computing the tariff reductions by Morocco and Tunisia is more complicated, as for these detailed schemes different time paths are foreseen. In the current analysis, we ignore differences in the speed of liberalisation undertaken by the two countries. We thus develop a scenario that removes all trade barriers on imports from the EU for all industrial tariff lines (at the 6-digit level) that are mentioned in Annexes 3 through 6 of the EuroMed Agreements. Table 5.1 presents the tariff reductions as computed from the EuroMed Agreements.

The EuroMed Agreements foresee an elimination of almost all tariffs on manufactured imports originating in the EU. This is reflected at the level of GTAP sectors in Table 5.1 by the (almost) complete elimination of tariffs on the manufacturing sectors. The EuroMed Agreements are very generic in character, subjecting more or less the same tariff lines to reductions in Morocco and Tunisia. Despite this generic character, Table 5.1 indicates rather different tariff reductions for several sectors (compare columns 1 and 2). The largest differences are found in the non-manufacturing sectors and are related to the manner in which tariffs at the 6-digit level are aggregated to the GTAP sector level. Several manufacturing tariff lines are linked to non-manufacturing GTAP sectors. For these sectors only a limited number of tariff lines are subject to reductions. To compute the tariff reductions applied in the GTAP model, we compute the tariff reduction at tariff-line level after which we aggregate again to the GTAP sector level using trade flows as weights. Having different trade flows, the resulting aggregate reductions at the GTAP sector level differ for Morocco and Tunisia. Basing tariff reductions on the detailed information available with version 6 of the GTAP database therefore results in differential tariff reductions, despite the reference to rather generic EuroMed Agreements.

Table 5.1 EuroMed Agreement scenario by sector and region (% reduction in tariffs)

Sector	MPC tariffs on imports originating in the EU		EU tariffs on imports originating in the MPCs	
	Morocco	Tunisia	Morocco	Tunisia
	(1)	**(2)**	**(3)**	**(4)**
1 Cereals	0.0	0.0	0.0	0.0
2 Oils seeds & vegetable oils	-11.6	-2.4	0.0	-0.2
3 Vegetables, fruit & nuts	0.0	0.0	0.0	0.0
4 Spices and other crops	-3.0	-11.5	-0.4	-17.7
5 Plant-based fibres	0.0	0.0	0.0	0.0
6 Sugar cane, sugar beet, sugar	0.0	0.0	0.0	0.0
7 Animal products and wool	-0.2	-0.4	0.0	0.0
8 Milk & dairy products	-1.9	-0.9	-0.1	0.0
9 Natural resource extraction	-22.6	-81.8	0.0	0.0
10 Food & beverages	-65.5	-58.8	-60.6	-77.2
11 Textiles & leather	-100.0	-99.7	0.0	0.0
12 Petro-chemicals	-100.0	-99.9	0.0	0.0
13 Wood, paper & mineral products	-100.0	-100.0	0.0	0.0
14 Metals and metal products	-100.0	-100.0	0.0	0.0
15 Various types of equipment	-95.1	-100.0	0.0	0.0
16 Motor vehicles & manufactures	-100.0	-84.4	0.0	0.0

Sources: Annexes 3 through 6 of the EuroMed Agreements, MAcMap and authors' computations.

The large number of zeros in columns 3 and 4 reflects the earlier removal of trade barriers that was implemented unilaterally in the 1970s by the EU. We computed the effects of a removal of tariffs on industrial goods (HS chapters 18 through 97) and on fishery products (chapter 3 and some additional 6-digit lines mentioned in protocol 2 of the EuroMed Agreements). As Table 5.1 shows, there are few tariffs remaining on trade in industrial goods. The only significant reduction in percentage terms is in food (61 and 77%). These remaining tariff barriers are probably owing to protection on the agricultural components of industrial goods as specified in Annex 1 of the EuroMed Agreements, which we cannot isolate because of lack of data. Regardless of the strong reduction in percentage terms, the reduction applies to an initial tariff of only 1.5% for Morocco and 1.6% for Tunisia, thus not granting much in terms of additional market access.

Overall the current agreements can be described as a significant lowering of trade barriers by Morocco and Tunisia (the percentages in Table 5.1 apply the levels depicted in Figures 5.1a and 5.1b), while their access to the EU market does not improve significantly.

The tariff reductions in Table 5.1 are used as input data for the scenario analysing the impact of the current EuroMed Agreements. Table 5.2 presents a summary description of the model scenarios employed in this study. The first set of scenarios refers to the current EuroMed Agreements (coded as EA). Given the restricted focus of the current EuroMed Agreements on industrial trade, the question arises as to what the impact would be if the agreements were to encompass agricultural trade as well. Such a full liberalisation (FL) of EU–Moroccan and EU–Tunisian trade is analysed with the second set of scenarios. Both contain two scenarios, differing in whether or not tariffs are replaced by a (consumption) tax. As we see when analysing the results, assumptions about the way in which MPC governments deal with the change in tariff revenues greatly affects the outcomes of the simulations.

The impact of the policy changes simulated in each of the scenarios is compared with a base run of the model that incorporates ongoing policy changes described in section 5.2. Tables 5.3a and 5.3b present the main indicators of the international trade relations of Morocco and Tunisia derived from the base run that serves as a reference point in our study. This provides a background for understanding the effects induced by the different policy changes.

Table 5.2 Description of model scenarios

Code	Description	Tariffs	Employment	Tariff replacement
EA_n	Current EuroMed Agreements with unemployment, no tariff replacement	See Table 5.1 for reduction percentages	Unemployment of skilled and unskilled labour	No replacement
EA_r	Current EuroMed Agreements with unemployment, with tariff replacement	See Table 5.1 for reduction percentages	Unemployment of skilled and unskilled labour	Tariffs replaced with consumption tax
FL_n	Full elimination of tariffs on EU–Moroccan and EU–Tunisian trade, no tariff replacement	All tariffs eliminated on EU–Moroccan and EU–Tunisian trade	Unemployment of skilled and unskilled labour	No replacement
FL_r	Full elimination of tariffs on EU–Moroccan and EU–Tunisian trade, with tariff replacement	All tariffs eliminated on EU–Moroccan and EU–Tunisian trade	Unemployment of skilled and unskilled labour	Tariffs replaced with consumption tax

Source: Authors' data.

Starting with the structure of the economies we find the services sector to dominate total GDP with a 57.6% share in Morocco and a 66.8% share in Tunisia. Note that in these cases services include the government (see the appendix to this chapter for the activities captured in each of the sectors). Agriculture plays an important role in both countries, accounting for 25.5% and 20.4% of GDP in Morocco and Tunisia, respectively. The role of agriculture in exports is less pronounced, and manufacturing plays an important part in earning foreign exchange, especially in textiles and leather (26.3% of Moroccan exports and 32.8% of Tunisian exports). Note also that the baseline construction includes the accession of China to the

WTO as well as the phasing-out of the Agreement on Textiles and Clothing, both of which lead to a contraction of the textiles and leather sector. The effects are most noticeable in Tunisia with a reduction of the textiles sector by 12.1% and a 1.9% loss of employment. In Morocco there is a much smaller contraction (0.7%) and the reallocation of production factors to other sectors increases employment by 0.2%. In spite of the changes in the global economy, textiles remain an important source of export earnings, even for Tunisia. Non-agricultural sectors dominate imports even more strongly, with the role of agricultural products limited to 22.4% in Morocco and 9.8% in Tunisia. Agricultural imports are clearly more important in Morocco than in Tunisia.

With regard to the trade balance the two countries have a striking similarity in qualitative terms. Both have a negative trade balance, with the value of imports exceeding exports by $415 million in Morocco and $715 million in Tunisia. The importance of the textiles sector in exports is reflected by its positive trade balance in both countries, being the only manufacturing sector with a positive contribution to net exports. Among the agricultural sectors only food and beverages, animal products and vegetables have a positive trade balance in both countries, while for Tunisia vegetable oils (olive oil) also shows positive net exports. The negative trade balances are further reflected in the limited number of sectors in which domestic production exceeds the total use of sector output (self-sufficiency).

The EuroMed Agreements are limited to liberalising trade with the EU. Their impact thus depends on the importance of the EU as a trading partner. The right panes of Tables 5.3a and 5.3b depict the share of exports or imports by sector traded with northern[3] and Mediterranean EU member states and Tunisia or Morocco.

---

[3] We added trade with the new EU member states from Eastern Europe to the northern EU member states. Given the very limited trade with the accession countries this does not influence the analysis.

Table 5.3a Basic international trade indicators for Morocco

| | General indicators | | | | | Importance of EuroMed destinations | | | | | |
| | | | | | | Northern EU member states | | Mediterranean EU member states | | Tunisia | |
	Share of GDP (%)	Share of exports (%)	Share of imports (%)	Trade balance ($ mn)	Self-sufficiency[a] (%)	Share exports (%)	Share imports (%)	Share exports (%)	Share imports (%)	Share exports (%)	Share imports (%)
Agricultural and mining sectors											
Cereals	5.6	0.2	5.3	-599	74	60.2	36.9	10.4	1.2	0.0	0.0
Oil seeds & vegetable oils	0.3	0.0	1.5	-179	54	30.7	7.7	25.2	8.6	0.0	0.7
Vegetables, fruit & nuts	4.6	4.2	0.4	428	127	73.2	21.2	4.0	15.9	0.0	17.1
Spices and other crops	0.3	0.6	1.3	-90	47	47.8	12.9	12.2	16.6	0.6	0.0
Plant-based fibres	0.6	0.1	0.4	-38	89	37.8	1.2	9.4	6.4	0.1	0.0
Sugar cane, sugar beet, sugar	0.8	0.1	0.5	-47	93	42.6	4.0	4.8	0.2	16.3	0.0
Animal products and wool	5.6	0.7	0.5	19	99	37.6	71.1	8.2	2.8	0.0	0.4
Milk & dairy products	1.5	0.2	1.0	-98	84	2.6	87.2	0.2	4.5	0.0	0.1
Natural resource extraction	4.0	5.4	9.6	-519	69	19.4	1.9	37.7	2.0	0.4	0.0
Food & beverages	2.1	8.0	1.8	697	119	20.9	32.3	35.4	16.1	0.5	0.5
Non-agricultural sectors											
Textiles & leather	5.8	26.3	16.6	1,035	102	70.2	59.2	19.8	27.6	0.3	0.1
Petro-chemicals	3.2	10.2	11.2	-153	90	27.7	40.0	11.8	23.4	1.1	1.2
Wood, paper & mineral products	3.7	3.0	6.5	-425	85	38.9	38.4	30.6	30.8	3.3	0.4
Metals and metal products	1.3	0.7	4.7	-480	72	23.6	35.6	38.0	35.5	5.0	0.5
Various types of equipment	1.8	10.8	22.8	-1,457	59	42.5	55.4	12.9	25.4	0.2	0.2
Motor vehicles & manufactures	1.0	1.2	4.8	-426	71	45.9	50.6	8.2	28.7	0.8	0.3
Services and activities nec	57.6	28.3	11.0	1,916	107	34.5	27.7	6.4	12.3	0.1	0.2
Total	100.0	100.0	100.0	-415	–	44.1	41.0	16.1	20.5	0.4	0.4

[a] Self-sufficiency is measured as the domestic share in total use.

Source: Authors' calculations – model simulation, base run.

Table 5.3b Basic international trade indicators for Tunisia

| | General indicators | | | | | Importance of EuroMed destinations | | | | | |
| | | | | | | Northern EU member states | | Mediterranean EU member states | | Morocco | |
	Share of GDP (%)	Share of exports (%)	Share of imports (%)	Trade balance ($ mn)	Self-sufficiency[a] (%)	Share exports (%)	Share imports (%)	Share exports (%)	Share imports (%)	Share exports (%)	Share imports (%)
Agricultural and mining sectors											
Cereals	0.7	0.3	3.6	-322	55	33.9	33.8	6.4	0.8	0.0	0.0
Oil seeds & vegetable oils	0.2	1.7	0.7	78	144	0.4	15.0	82.1	20.1	0.8	0.0
Vegetables, fruit & nuts	7.8	1.1	0.3	73	102	55.0	34.4	17.6	22.1	8.1	0.5
Spices and other crops	0.2	0.1	0.8	-66	30	56.0	8.0	11.6	28.0	0.3	0.5
Plant-based fibres	0.0	0.1	0.4	-36	21	32.2	1.5	7.1	34.4	0.2	0.0
Sugar cane, sugar beet, sugar	0.3	0.0	0.2	-22	88	7.6	9.1	0.8	2.3	0.0	9.4
Animal products and wool	2.9	0.6	0.3	24	99	31.1	67.3	40.0	9.4	0.5	0.0
Milk & dairy products	0.1	0.1	0.2	-14	89	5.5	65.5	0.4	5.4	1.8	0.0
Natural resource extraction	6.3	4.2	1.5	230	109	26.8	6.0	57.6	17.2	0.0	1.6
Food & beverages	1.9	2.1	1.7	28	93	19.0	20.3	33.1	9.6	0.5	3.0
Non-agricultural sectors											
Textiles & leather	5.3	32.8	21.0	906	110	68.7	52.7	25.9	39.3	0.1	0.4
Petro-chemicals	2.5	8.9	14.0	-559	77	20.6	46.9	16.7	31.8	2.1	1.0
Wood, paper & mineral products	2.9	3.1	7.1	-411	75	35.6	48.7	19.1	26.0	1.0	1.6
Metals and metal products	0.3	1.8	4.9	-315	53	40.1	42.5	28.6	29.3	1.8	0.8
Various types of equipment	1.2	13.7	26.2	-1,305	50	72.5	58.9	14.2	22.1	0.4	0.1
Motor vehicles & manufactures	0.7	2.7	7.6	-488	52	58.0	60.7	17.0	26.0	0.6	0.2
Services and activities nec	66.8	26.8	9.4	1,486	111	36.0	30.8	6.6	9.2	0.1	0.2
Total	100.0	100.0	100.0	-715	–	39.1	39.6	15.9	21.0	0.4	0.4

[a] Self-sufficiency is measured as the domestic share in total use.

Source: Authors' calculations – model simulation, base run.

Analysing the differential importance of trade flows by trading partner provides some initial insight into the political economy of the EuroMed Agreements. Taken together, the EU accounts for 60.2% (44.1 and 16.1%) of Morocco's exports and 55.0% (39.1 and 15.9%) of Tunisia's exports. Comparable large percentages are found for EU imports to Morocco (60.2%) and Tunisia (60.6%). The EU is thus the major trading partner of both Morocco and Tunisia.[4]

We argue that the lack of progress in liberalising agricultural trade flows can be traced to the varying interests of northern and Mediterranean EU member states. Some indications of these interests can be found in Tables 5.3a and 5.3b. Although the Mediterranean EU member states are much smaller (their combined GDP is about a third of the GDP of the northern EU region), for some sectors they account for a very large share of exports. Most notable is the trade of vegetable oil, with 25.2% of Moroccan exports and no less than 82.1% of Tunisian exports being destined for Mediterranean EU countries. The picture in terms of vegetables is less clear, probably owing to the aggregated representation in the model. The exports to the northern EU are about three times as large as to the Mediterranean EU region, which corresponds to the difference in their GDPs. On the import side, a clear diverging interest is found for the animal products and dairy sectors, with 71.1 and 87.2% of Moroccan imports and 67.3 and 65.5% of Tunisian imports originating in the northern EU. For cereals we find a similar dominance of northern EU countries (although less pronounced because of imports from the US). Imports from the Mediterranean EU countries are limited for these 'temperate' agricultural products.

Despite its aggregated character the model does capture the main diverging interests of northern and Mediterranean EU member states: northern EU countries would benefit from improved access to North African markets for animal products, dairy and cereals, whereas Mediterranean EU countries would face a further increase in imports of vegetable oils. We do not find a similar pattern for vegetables, because of the aggregated nature of the vegetable sectors in the GTAP, which unfortunately combines Mediterranean products with more temperate

---

[4] These numbers reflect projected trade flows given the ongoing policy changes and therefore differ from the numbers based on past trade flows in the previous chapter.

vegetables and fruit. The data also reflect the importance of the EU as a trading partner of both Morocco and Tunisia. The EuroMed Agreements can thus be expected to have a significant impact on both economies.

5.4 Results of the current EuroMed Agreements

This section analyses the results of the current EuroMed Agreements. We focus on the current agreements in the first instance since these are already signed. The next sub-section compares the impact of expanding trade liberalisation to include agriculture, which could be achieved if the recently proposed negative-list approach to liberalising agricultural trade in the Mediterranean were to be adopted.

5.4.1 Income effects of the EuroMed Agreements without and with tariff replacement

Our analysis of the income effects of policy changes focuses on the equivalent variation (EV), a concise measure of the macroeconomic consequences of a (policy) change. The EV measures the change in income equivalent to the proposed policy change. More specifically, it measures the amount of income that should be given to (or taken away from) a household to attain a welfare equivalent to the welfare occurring after a (policy) change comes into effect. If a policy change results in a positive EV, this number represents the additional income that could be generated if the policy were implemented. If total EV is positive the winners could potentially compensate the losers. Apart from this general conclusion on the potential for compensation the EV does not consider distributive issues.

We start by analysing the effect of implementing the EuroMed Agreements without tax replacement (left pane of Table 5.4). The elimination of tariffs on trade with the EU has two major effects: i) reducing distortions in the economies of mainly Morocco and Tunisia; and ii) diverting trade from the rest of the world (ROW) to the EU. The first effect is strongest for Morocco, which initially levies higher tariffs on European manufactured goods than Tunisia (see the discussion of tariff profiles with Figures 5.1a and 5.1b). In absolute terms the gains for Morocco are highest (an annually recurring $2,692 million or 7.92% of GDP in the base run). Since northern EU member states account for a larger share of trade (in both Morocco and Tunisia), they gain more from the lowering of tariff barriers ($823 million). Because of the limited economic importance of Morocco and Tunisia in the total trade of the European Union, the impact assessed in terms of base GDP remains an insignificant 0.01% for northern EU countries and 0.02% for those in the Mediterranean EU.

Table 5.4 Income effects of the current EuroMed Agreements (million US$, 2001)

	No tax replacement (EA_n)			With tax replacement (EA_r)		
	Total	Share of base GDP	Share of total gains	Total	Share of base GDP	Share of total gains
		(%)	(%)		(%)	(%)
Morocco	2,692	7.92	65.76	-1,455	-4.28	82.19
Tunisia	907	4.67	22.15	-520	-2.68	29.37
Northern EU	823	0.01	2.10	619	0.01	-34.94
Mediterranean EU	371	0.02	9.07	282	0.01	-15.92
ROW	-699	0.00	-17.07	-696	0.00	39.30
Total	4,094	0.01	100.00	-1,771	-0.01	100.00

Note: Computed from EV.

Source: Authors' calculations based on model simulations.

The lowering of barriers on trade with the EU gives European producers an edge on producers in the rest of the world. The EU already has lower tariffs on manufactured goods (see Figures 5.1a and 5.1b) and the EuroMed Agreements further increase this advantage. This results in a decrease of trade with the ROW, shown in Table 5.4 as a welfare loss for the ROW of $699 million.

If we then turn to the right pane in Table 5.4 we find that the results for a scenario with tax replacement (EA_r) give a radically different picture of the income effects of the EuroMed Agreements. Although the effects for the EU are comparable to the first scenario we now find welfare declining in both Morocco ($-1,455 million or -4.28% of base GDP) and Tunisia ($-520 million or -2.68% of base GDP). This radical change in the assessment of the agreements is driven by a different assumption on the need for replacing the tariff revenues. The second scenario assumes that the government fully replaces tariff revenues by a consumption tax to sustain its expenditures. This assumption implies that the distortions of an import tax are replaced by a distorting consumption tax. As indicated by the total income loss of $1,771 million, the distortions of the consumption tax exceed the reduced distortions of the import tax.

The strong impact of tax replacement, turning significant gains into significant welfare losses, comes rather unexpectedly. Both countries are small, which implies that changes in domestic demand would not affect international prices. Consequently, import taxes only raise domestic prices, not international prices.[5] The tax replacement then only shifts the manner in which taxes are collected from an import to a consumption tax, and one would not expect such dramatic shifts in welfare.

The strong impact is the result of three interrelated features of our modelling exercise: a uniform shift in taxes, the Armington assumption and the unemployment closure. We use a change in closure to model tax replacement – we fix the ratio of tax revenues to national income and use an endogenous tax rate on private consumption to compensate for the shortfall of import tariff revenues. The tax shift leads to a significant rate of taxation on consumption of 9.5% in Morocco and 8.5% in Tunisia, which exacerbates existing distortions. This effect is visible in the different repercussions on Morocco and Tunisia. Tunisia has little variation in consumption taxes, while in Morocco the existing and possibly inefficient dispersion of tax rates (with much higher tariffs on manufactured goods) is aggravated by the shift in taxes. Making domestic taxes more uniform may increase the benefits from trade liberalisation, but the coverage of tax data in our database is not sufficient to allow a satisfactory analysis of this fiscal issue.

Given the uniform tax rates in Tunisia the puzzle remains as to how the change in taxes has such a strong effect on welfare. In a recent paper, Taylor & von Arnim (2006, pp. 23-26) show that when modelling bilateral trade flows using the Armington assumption (as in virtually all applied CGE models, including the GTAP), reducing tariffs reduces demand and welfare gains if government expenditures are fixed and tariff revenues are replaced by a consumption tax. The intuition of this result is that consumption is reduced by the rise in tax rates. The size of this reduction depends on the share of imports in the Armington aggregate, which will always be less than one (since some domestic goods are consumed as well). By replacing tariff revenues with consumption taxes consumers bear the

---

[5] Using the Armington assumptions, all goods are region-specific, which gives even small countries a limited amount of market power. For all practical purposes we can maintain the small-country assumption when analysing results.

full burden of the increased taxes: the incidence of the tax falls on a much broader base. The net effect on consumption is therefore always negative. Morocco has initially higher tariffs on manufactured goods and thus the negative effect on consumption will be larger for Morocco, contributing to its higher welfare losses.

In our model the welfare losses through the consumption side are multiplied by the assumption of unemployment. A reduction in consumption will result in a contraction of production, which will increase unemployment and further reduce consumption. This multiplier effect is considerable. A simulation of the implementation of the agreements with tax replacement but assuming full employment results in much smaller welfare losses ($107 million for Morocco and $58 million for Tunisia). The unemployment closure thus strongly interacts with the Armington fiscal effect, but negative welfare effects remain even with full employment.

The combination of distorting initial consumption taxes, imperfect substitution of domestic and foreign goods in consumption (Armington) and the multiplier effect of unemployment generate large negative welfare effects of tariff reductions with revenue replacement. These findings underscore the need to carefully address the budgetary implications of a tariff reduction while accounting for the overall economic setting (especially unemployment). Such an assessment requires detailed modelling of the tax system and its effects on the economy, for which the GTAP with its rather limited data on government revenues and taxes is less suited.

Next to overall income effects employment is a major concern in both Morocco and Tunisia. The changes in employment are closely linked to the overall income effects, with employment increasing in the first scenario and decreasing in the case of tariff replacement (Table 5.5). With a single representative household we cannot address issues of income distribution across households. Changes in factor prices, however, do provide some initial insight into the distributive effects of the agreements. Poor households tend to have only unskilled labour and possibly some land (if located in rural areas).

The current EuroMed Agreements only entail changes in import tariffs on manufactured goods, which do not use land. The changes in returns to land in Table 5.5 indicate that the agricultural sectors are indirectly affected by the tariff reductions. Without tax replacement the improved allocative efficiency raises incomes, stimulating demand for (agricultural) consumption goods. Increasing the consumption tax to replace tariffs reduces demand, which translates into a reduced return to land. With real wages fixed the price of labour cannot accommodate and all the adjustment has to occur in the amount of labour employed. In contrast, there is limited scope for quantity adjustments for the immobile production factor of land and most adjustments occur through changes in its rental rate. Capital also witnesses drops in returns, for many of the same reasons. Although capital is assumed to be inter-sectorally mobile, the overall contraction of the economy provides only limited opportunities for the reallocation of capital across sectors.

Table 5.5 Employment and factor prices with the EuroMed Agreements (% change to base)

	No tax replacement (EA_n)		With tax replacement (EA_r)	
	Morocco	**Tunisia**	**Morocco**	**Tunisia**
Employment				
Unskilled labour	14.9	11.1	-6.8	-5.1
Skilled labour	15.6	11.7	-8.3	-5.9
Factor prices				
Land	24.5	11.8	-14.2	-19.2
Unskilled labour	0.0	0.0	0.0	0.0
Skilled labour	0.0	0.0	0.0	0.0
Capital	12.0	9.6	-7.0	-4.2

Source: Authors' calculation based on model simulations.

In terms of income distribution this implies that not only are poor households in urban areas affected by the agreements, poor households in rural areas are as well. General equilibrium effects transmit the policy changes in the manufacturing sector to rural areas by changing the demand for agricultural goods. Rural wage labourers and small farmers thus benefit from the reduced import tariffs while being harmed by the consumption tax.

5.4.2   *Drivers of the aggregate income effects*

We found considerable differences in the income effects of the EuroMed Agreements, depending on whether or not tariffs are replaced by a consumption tax. Some general explanations of the underlying mechanisms have been discussed. We now examine in more detail the adjustments induced by the agreements.

The EV on which Table 5.4 is based provides a summary of statistics concerning the total effect of different drivers of welfare changes. Table 5.6 presents the contribution of four main drivers to the total income gains expressed in terms of base GDP: allocative efficiency, employment and terms of trade.

The **allocative efficiency** effects relate to distortions induced by taxes. The removal of import tariffs reduces distortions, allowing factors of production (land, labour and capital) to move to their most efficient use. Increased efficiency translates into lower prices, promoting the expansion of supply and demand. Increasing the consumer tax to replace tariff revenues aggravates distortions working in the opposite direction of the import tariff reductions.

Focusing on the top part of Table 5.6 we find large allocative efficiency effects for Morocco (30% of net gains) and Tunisia (44% of net gains). Contributions to the EU regions are minimal, reflecting the low initial tariffs and limited importance of imports from Morocco and Tunisia. With Morocco and Tunisia having the highest initial tariffs and the EU being a major trading partner, they are also benefiting most from removal of the tariffs. The flip side is that they are also affected most by the distortions induced by an increased consumption tax replacing tariff revenues (see the lower half of Table 5.6). With tax replacement there are two opposing forces: the removal of import tariffs improves efficiency while the increase in consumption taxes reduces efficiency. A net efficiency loss remains for both countries. The net loss is more important in Tunisia (24% of total loss) than in Morocco (4%). This difference stems from efficiency gains in Morocco's manufacturing sectors, which are related to higher initial tariffs on manufactured imports in Morocco (see the discussions with Figures 5.1a and 5.1b) and which create more scope for improving efficiency by reducing tariffs than in Tunisia.

The **employment effect** is the most important factor determining the net income effects, both in terms of gains with only tariff removal (88% for Morocco and 77% for Tunisia) and in terms of losses (76% for Morocco and 64% for Tunisia). The large role of changes in employment is a direct result of the modelling of the labour market. In the unemployment closure we fix the real wage while allowing the number employed in the economy to match demand. This implies that sectors can expand production without facing an increase in production costs, thus increasing their competitiveness. The additional labour market entrants in turn increase consumption demand, stimulating production further. A contraction of the economy has a similarly strong but opposite effect: shrinking employment reduces demand, further contracting production.

Table 5.6 Income effects of the EuroMed Agreements by source

	Share of base GDP	Source (% of gain or loss)			
	(%)	Allocative efficiency	Employ-ment	Terms of trade*	Total
No tax replacement (EA_n)					
Morocco	7.92	30	88	-18	100
Tunisia	4.67	44	77	-21	100
Northern EU	0.01	3	0	98	100
Mediterranean EU	0.02	3	0	97	100
ROW	0.00	19	9	72	100
With tax replacement (EA_r)					
Morocco	-4.28	4	79	17	100
Tunisia	-2.68	24	64	12	100
Northern EU	0.01	5	0	95	100
Mediterranean EU	0.01	2	0	98	100
ROW	0.00	12	10	78	100

* This column includes small terms-of-trade effects on the balance of payments account through changes in the prices of investment and savings.

Source: Authors' calculations based on model simulations.

The changes in tariffs and in consumption tax affect sectors differently. Table 5.7 presents the changes in demand for labour by sector. The general equilibrium effects of the reduction in import tariffs on manufactured goods are clearly indicated by the changes in demand for labour in agricultural sectors when only tariffs on manufactured goods are removed (see the left pane of Table 5.7). Increased employment in the manufacturing sectors is transmitted to agricultural sectors by an increased demand for consumption goods. The agricultural sectors are thus not shielded from the current EuroMed Agreements, despite their focus on manufactured goods.

Table 5.7 Demand for labour by sector with the EuroMed Agreements (% change)

	No tax replacement (EA_n)		With tax replacement (EA_r)	
	Morocco	Tunisia	Morocco	Tunisia
Agricultural sectors				
Cereals	5	1	-9	-10
Oil seeds & vegetable oils	8	-7	-11	-18
Vegetables, fruit & nuts	3	3	-7	-6
Spices and other crops	-4	-7	-14	-18
Plant-based fibres	9	7	-7	0
Sugar cane, sugar beet, sugar	11	8	-9	-8
Animal products and wool	22	8	0	-6
Milk & dairy products	9	9	-12	-8
Natural resource extraction	-3	-1	-8	-5
Food & beverages	11	7	-7	-10
Non-agricultural sectors				
Textiles & leather	53	46	23	24
Petro-chemicals	4	9	-18	-9
Wood, paper & mineral products	-2	2	-24	-17
Metals and metal products	0	7	-23	-8
Various types of equipment	18	9	-7	-11
Motor vehicles & manufactures	-2	7	-23	-13
Services	15	11	-9	-6

Source: Authors' calculations based on model simulations.

The textiles sector exhibits the largest increase in demand for labour. Note that there is no change in tariffs faced by textile exports to the EU since these were eliminated in the 1970s (see the changes in tariffs in Table 5.1). Also note that the increases reported in Table 5.7 are relative to the base data, which included a contraction of textiles following China's WTO accession and abolition of the Agreement on Textiles and Clothing. Textiles benefit from a rationalisation of the manufacturing sectors when the protection from European imports is removed. As a result factors of production, including labour, move to their most efficient use. This increase is strong enough to undo the contraction of textiles captured in the baseline.

Overall the impact of the EuroMed Agreements on employment in manufacturing is not as devastating as one might have expected given their relatively high initial protection, especially in Morocco. There are two major explanations for this outcome. First, tariffs on imported intermediate inputs are removed as well, which reduces production costs in sectors using European imports in their production. Second, by accounting for unemployment we fixed the real wage rate. This implies that labour becomes relatively more attractive compared with other factors of production for which prices rise (see Table 5.5). Sectors thus adjust their input mix and employ more labour, which contributes to increased employment.

As discussed above, the presence of unemployment may also lead to a stronger contraction of the economy. This is reflected in the large reductions in employment in the right pane of Table 5.7, presenting results for the scenario in which tariffs are replaced. The only major exception is the textiles sector, which keeps its increased demand for labour. Our reference point is a base run that includes the phasing-out of the Agreement on Textiles and Clothing as well as the WTO accession of China, a major competitor in the textiles market. Despite these ongoing changes serious distortions in the global textiles market persist, implying that the preferential access to the European textiles market remains important for Morocco and Tunisia. A side effect of the remaining protection on agricultural production is that in Morocco employment increases in the animal products sector, even when consumption taxes are increased. This sector, however, is the most protected in Morocco (see Figure 5.1a). This shift of labour to such a highly protected sector may increase the costs of a future agreement liberalising agricultural trade.

Returning to Table 5.6 we find that the third major driver of changes in income is the **terms-of-trade effect**. The terms of trade are a summary measure indicating the ratio of prices received for exports to prices paid for imports. We find declining terms of trade for Morocco and Tunisia in both scenarios, accounting for 12 to 19% of the income changes. The terms-of-trade effect is a macroeconomic phenomenon related to the balance of payments. The balance of payments measures the inflow of money from exports and investments and the outflow of money through imports and savings. Although some adjustments occur in savings and investments, the major adjustments occur in imports and exports.

By removing import tariffs the EuroMed Agreements increase imports in Morocco and Tunisia. In order to maintain the balance of payments, exports need to rise to generate the foreign exchange needed for the increased imports. Growth in exports requires the exports of Morocco and Tunisia to become cheaper than their competitors' products. The main mechanism to achieve this adjustment is normally a reduction in the costs of production factors in order to reduce the prices of exports. If we look at Table 5.5, however, we find that in the cases of Morocco and Tunisia factor prices increase when tariffs are not replaced. The origin of their relatively cheaper exports thus lies not in lower factor prices, but in the lower costs of imported intermediate inputs. The drop in prices of intermediate inputs allows an expansion of exports by reducing relative prices despite a rise in factor prices. This is an illustration of the so-called 'Lerner symmetry': shielding the economy from imports ultimately reduces exports. Reducing import protection boosts exports while allowing incomes to rise.

5.4.3 A summary of the impact of the current EuroMed Agreements

Summarising we find that the current EuroMed Agreements have the potential to benefit Morocco and Tunisia. The direction of the impact, however, strongly depends on domestic policies. If tariff revenues are replaced by a consumption tax the income effects are negative, unemployment increases and returns to land decrease. For poor households relying on labour and possibly land (in the case of the rural poor) changes in domestic policies following trade liberalisation are thus essential.

When delving a bit deeper into the drivers of the income changes we find that accounting for unemployment plays a central role in the overall assessment of the agreements. Unemployment strongly enhances both the positive and negative consequences of policy changes. Analysing the allocative efficiency effects we find the textiles sector to be expanding in both scenarios. This result appears to be related to the remaining distortions in the global textiles trade. These distortions imply that the preferential access to European markets remains valuable for Morocco and Tunisia. Analysing the terms-of-trade effects we find that the removal of import barriers allows Morocco and Tunisia to increase the competitiveness of their exports through reduced costs of imported intermediate inputs. This development in turn allows factors of production and thus households to benefit from increased factor prices.

In this section we have focused on the current EuroMed Agreements, which are limited to manufacturing. A natural question then arises as to the impact of including agricultural sectors in the trade agreements, especially given the recent proposal for a negative-list approach to liberalising agricultural trade. There appear to be two key issues related to the political economy of further reforms: i) contradictory EU interests and ii) the future costs of liberalisation. Opposite gains by the northern and Mediterranean EU member states seem to be a main factor in the current virtual absence of agricultural reform in the agreements. The above analysis furthermore indicates an expansion of the heavily-protected animal products sector in Morocco. This expansion suggests that the current one-sided agreements may increase the future costs of liberalising agricultural trade.

5.5   Results when agriculture is incorporated

The opposite interests of the northern and Mediterranean EU regions will come into play when agricultural trade is liberalised. Lowering tariff barriers on agricultural trade is also expected to affect the most strongly protected sectors of Morocco, which would expand its employment with the implementation of the current EuroMed Agreements. Again we analyse full liberalisation scenarios without tariff replacement (FL_n) and with tariff replacement by a consumption tax (FL_r). In the scenarios we assume all tariffs are removed, i.e. we analyse a full liberalisation of trade with the EU for Morocco and Tunisia.

The additional gains from liberalising agricultural trade differ for Morocco and Tunisia (Figure 5.2). In the absence of tariff replacement, Morocco gains an additional 2.2 percentage points to its base GDP whereas Tunisia gains an additional 6.9 percentage points. In terms of the share of base GDP the gains for Tunisia are larger here than for Morocco, which gained more from the liberalisation of manufactured trade.

In the case of tariff replacement the inclusion of agricultural trade does not yield much difference in the overall impact for Morocco; its income loss becomes slightly deeper by an additional 0.1 percentage points.

For Tunisia the inclusion of agriculture yields much more, even reversing the loss of income and producing an income gain of 1.0% of base GDP.

Figure 5.2 Income effects for Morocco and Tunisia by scenario (% of base GDP)

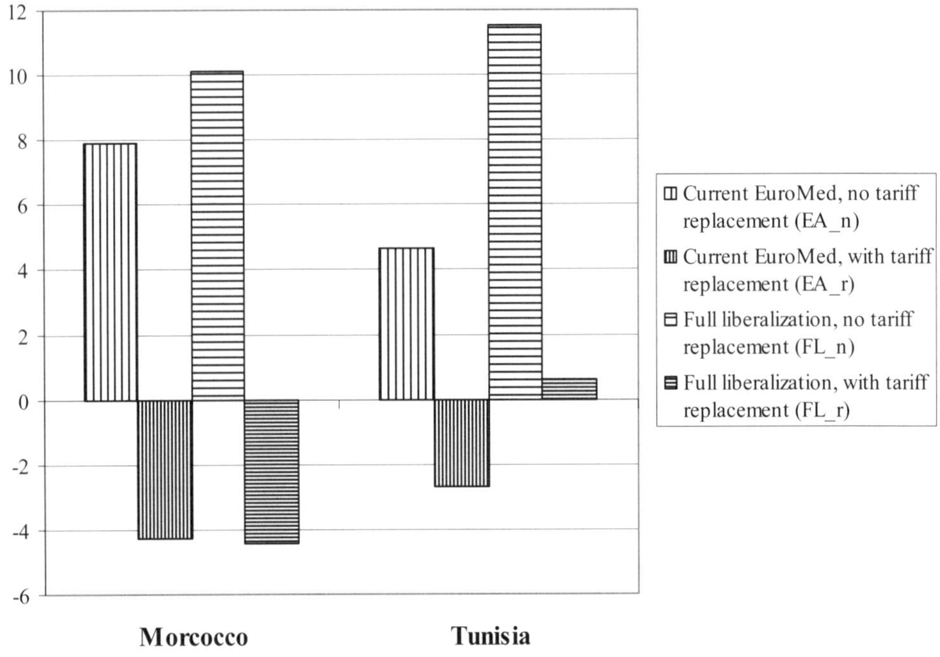

Source: Authors' calculations based on model simulations.

The differential impact of including agriculture in the agreements can in part be explained by differences in initial tariffs (see Figures 5.1a and 5.1b). Tunisia has higher initial agricultural tariffs and therefore stands to gain more from a liberalisation of its agricultural trade in terms of improved allocative efficiency. The second important factor is a spectacular increase in vegetable oil production when access to EU agricultural markets improves. In the base run the majority of these exports (82.1%) are to the Mediterranean EU regions (see Table 5.3b). This outcome implies that the Mediterranean regions are also affected most by the liberalisation of agricultural trade.

The changes in production in the two full liberalisation scenarios clearly indicate that the growth in vegetable oil production in Tunisia drives the aggregated effects (Table 5.8).

Table 5.8 Output by sector and region with full liberalisation (% change)

	No tax replacement (FL_n)				With tax replacement (FL_r)			
	Morocco	Tunisia	EU North	EU Med.	Morocco	Tunisia	EU North	EU Med.
Agriculture	7	16	0	0	-6	6	0	0
Cereals	-6	-44	3	1	-18	-51	2	1
Vegetable oils	10	856	0	-7	-7	795	0	-6
Vegetables & fruit	11	15	0	0	2	5	0	0
Spices & other crops	-5	-20	0	1	-13	-29	0	0
Plant-based fibres	10	0	-1	0	-6	-8	0	0
Sugar	11	11	0	0	-6	-3	0	0
Animal products	21	1	0	0	2	-12	0	0
Dairy	-18	-18	0	0	-31	-30	0	0
Food & beverages	11	0	0	0	-3	-10	0	0
Natural resources	-2	-3	0	0	-6	-6	0	0
Manufacturing	15	8	0	0	-2	-4	0	0
Textiles & leather	54	26	0	0	31	15	0	0
Petro-chemicals	-1	-2	0	0	-13	-11	0	0
Wood, paper	-6	-6	0	0	-20	-18	0	0
Metal products	-3	1	0	0	-19	-13	0	0
Equipment	17	-3	0	0	-3	-20	0	0
Motor vehicles	-5	-3	0	0	-20	-15	0	0
Services	11	11	0	0	-5	-2	0	0

Source: Authors' calculations based on model simulations.

Whether or not tariffs are replaced, vegetable oil production shows an extraordinary eight-fold increase. This increase negatively affects the Mediterranean EU regions, although compared with the remarkable changes in Tunisia the effects are rather modest (-7% and -6% with tariff replacement). In both scenarios Morocco and Tunisia reduce the production of cereals, benefiting both the northern and Mediterranean EU regions. Apart from a 1% decline in plant-based fibres in the northern EU there are no noticeable effects on European agriculture. This result illustrates the limited economic importance of Morocco and Tunisia for the EU as a whole.

Based on the high initial protection on animal products in Morocco we expected a decline following trade liberalisation. Apparently animal production is more competitive than expected, expanding in both scenarios (21% and 2% with tariff replacement). The labour that moved from manufacturing to animal production with the implementation of the current EuroMed Agreements (Table 5.7) thus does not have to move again when agricultural trade is liberalised. Textiles and leather also maintain the expansion found with only liberalising manufacturing trade. The major contraction of cereals and dairy, however, does entail a loss of unemployment. In the case of Morocco this results in a slightly higher loss of employment when liberalisation is expanded to include the agricultural sector and tariffs are replaced (Figure 5.3).

Figure 5.3 Employment for Morocco and Tunisia by scenario (% change)

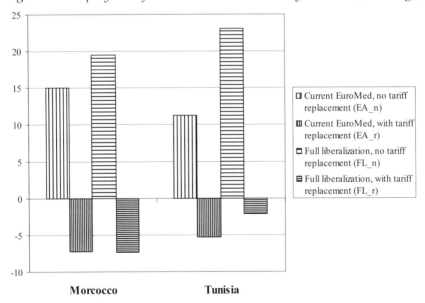

Source: Authors' calculations based on model simulations.

Conclusions

This study analyses the economic effects of the EuroMed Agreements using a multi-sectoral, multi-regional model. Our numerical analysis, which focuses on Morocco and Tunisia, is built from a detailed commodity profile (6-digit level) of the provisions of the EuroMed Agreements. These current agreements turn out to be very asymmetrical: the EU essentially does not give any new concessions (maintaining its protection of agricultural sectors), while the MPCs will have to open their manufacturing markets to European competitors.

Because of the small size of the Moroccan and Tunisian economies, improved access to their markets for manufactured goods yields insignificant benefits for the EU – a mere 0.01% of base GDP for the northern EU region and 0.02% for the Mediterranean EU. Being located closer to Morocco and Tunisia, the Mediterranean EU countries trade more with them and therefore gain more from the current agreements.

The impact of the current EuroMed Agreements on Morocco and Tunisia is found to depend on their need to replace reduced tariff revenues with an increased consumption tax. If such replacement is not needed, both economies benefit from an improved allocation of factors of production with the removal of distorting import tariffs. Their economies expand, increasing both employment and factor earnings. A rather different picture emerges when tariff revenues are replaced by a consumption tax. This consumption tax aggravates distortions in the economy, and for some commodities implies a shift from subsidising to taxing consumption. Apart from the political issues related to the implementation of such a tax increase, we find a decline in income and a loss of employment. This unexpected and strongly negative impact of tariff replacement (reversing the outcomes of the agreements) can be attributed to a combination of the tax instrument chosen (a uniform shift in consumption taxes), the fiscal effects of the Armington assumption used to model bilateral trade and the assumption of unemployment. From an analytical perspective, this highlights the need to model in detail how taxes affect the economy in order to design the change in tax policy to avoid negative consequences on incomes and employment.

The virtual absence of agricultural liberalisation in the current agreements can be put down to the diverging interests of northern and Mediterranean EU member states. Analysing a full liberalisation of all trade (including agriculture) with Morocco and Tunisia, we find these diverging interests reflected in the changes in output. The northern EU region benefits from an increase in cereal exports (as does, to a lesser extent, the Mediterranean EU region). The Mediterranean EU region faces a 7% reduction in its vegetable oil production owing to an eight-fold production increase in Tunisia. Despite this reduction the overall impact on the two EU regions remains insignificant, because of the limited size of the Moroccan and Tunisian economies.

Because of its initially higher level of agricultural protection Tunisia benefits more from an expansion of the current agreements to include agriculture. In the case of Morocco, the additional benefits (and costs, if tariffs are replaced) are limited. Driven by the increase in vegetable oil production Tunisia maintains its income gain even when tariffs are replaced. In terms of employment, however, both countries face a reduction in employment when tariffs need to be replaced, even when agriculture is included.

In conclusion we find that a liberalisation of EU agricultural production only has a significant impact on cereals (positive) and vegetable oil (negative for the Mediterranean EU countries). This latter finding is driven by an unrealistic eight-fold increase in Tunisian production. Given biophysical limitations on the expansion of olive oil (the main vegetable oil produced in Tunisia) the large competitive advantage identified in the model simulations are not likely to materialise when trade barriers are removed. This view implies that the negative impact on the Mediterranean EU region will be less than suggested by the model simulations, which are already found to be insignificant when related to GDP. More importantly, the sizable positive impact driving an overall positive income effect for Tunisia when agricultural trade is liberalised will not materialise.

We find the impact of the EuroMed Agreements on incomes and employment to hinge upon the domestic policies of Morocco and Tunisia in response to a loss of tariff revenues. If government expenditures can be reduced such that a replacement of tariffs is not required, both countries will benefit from increased factor incomes and employment. If tariff replacement is needed a careful consideration of the economy-wide impact is necessary to reduce the income and employment losses we find when uniformly increasing consumption taxes. This observation highlights the importance of embarking on domestic tax reforms in parallel with trade reforms.

The results in this study underscore that (apart from specific sector interests) the EuroMed Agreements are not of economic significance to northern or Mediterranean EU member states. Thus there is room to focus on an implementation of the agreements in sync with domestic policy reforms in the MPCs, to reap the potential benefits of the agreements to enhance growth and social stability at the southern borders of the EU. It further affords the possibility to avoid the potential negative implications when reduced tariff revenues need to be recovered elsewhere in the MPC economies.

References

Bulmer, E.R. (2000), *Rationalizing Public Sector Employment in the MENA Region*, Working Paper Series No. 19, World Bank, Washington D.C.

Dimaranan, B.V. and R.A. McDougall (eds) (2005), *Global Trade, Assistance, and Production: The GTAP 6 Data Base*, Center for Global Trade Analysis, Purdue University.

European Council (2003), *A Secure Europe in a Better World*, European Security Strategy presented by Javier Solana, Brussels, 12 December.

Garcia-Alvarez-Coque, J.-M. (2002), "Agricultural trade and the Barcelona Process: Is full liberalization possible?", *European Review of Agricultural Economics*, Vol. 29, No. 3, pp. 399-422.

Hertel, T.W. (ed.) (1997), *Global Trade Analysis: Modelling and Applications*, Cambridge University Press, Cambridge, MA.

Kuiper, M. (2004), *Fifty Ways to Leave your Protection, Comparing Applied Models of the Euro-Mediterranean Association Agreements*, ENARPRI Working Paper No. 6, CEPS, Brussels (retrieved from http://www.enarpri.org/publications.htm).

Taylor, L. and R. von Arnim (2006), *Modelling the Impact of Trade Liberalisation: A Critique of Computable General Equilibrium Models*, Oxfam International Research Report, Oxfam, Oxford.

Wallström, M. (2005), "The European Neighbourhood Policy and the Euro-Mediterranean partnership", SPEECH/05/171, delivered at the Euro-Mediterranean Parliamentary Assembly on 14 March in Cairo, European Commission, Brussels (retrieved from http://europa.eu/external_relations/news/ferrero/2005/ip05_42.htm).

Appendix: Aggregation schemes

Table A5.1 Aggregation of regions

Aggregate region	Code	GTAP V6 regions
1) Morocco	MOR	Morocco
2) Tunisia	TUN	Tunisia
3) Rest of North Africa	RNA	Rest of North Africa
4) Middle East	MEAST	Rest of Middle East
5) Northern EU members	EU_N	Austria
		Belgium
		Denmark
		Finland
		France
		Germany
		United Kingdom
		Ireland
		Luxembourg
		Netherlands
		Sweden
6) Mediterranean EU members	EU_M	Greece
		Italy
		Portugal
		Spain
7) EU accession countries	EU_A	Bulgaria
		Cyprus
		Czech Republic
		Hungary
		Malta
		Poland
		Romania
		Slovakia
		Slovenia
		Estonia
		Latvia
		Lithuania
8) US	USA	United States
9) All other regions	ROW	The remaining 55 GTAP regions

Table A5.2 Aggregation of sectors

Sector	Code	Sector	Code
1) Cereals	cereal	Wearing apparel (wap)	
Wheat (wht)		Leather products (lea)	
Cereal grains nec (gro)		12) Petro-chemicals	petchem
Processed rice (pcr)		Petroleum, coal products (p_c)	
2) Oil seeds & vegetable oils	oilcrp	Chemical, rubber, plastic prods (crp)	
Oil seeds (osd)		13) Wood, paper & mineral products	wd_min
Vegetable oils and fats (vol)		Wood products (lum)	
3) Vegetables, fruit & nuts	veg_frt	Paper products, publishing (ppp)	
Vegetables, fruit, nuts (v_f)		Mineral products nec (nmm)	
4) Spices and other crops	spices	Metals nec (nfm)	
Crops nec (ocr)		14) Metals and metal products	metal
5) Plant-based fibres	fibercrp	Ferrous metals (i_s)	
Plant-based fibres (pfb)		Metal products (fmp)	
6) Sugar cane, sugar beet, sugar	sugar	15) Various types of equipment	equip
Sugar cane, sugar beet (c_b)		Transport equipment nec (otn)	
Sugar (sgr)		Electronic equipment (ele)	
7) Animal products and wool	anim	Machinery and equipment nec (ome)	
Cattle, sheep, goats, horses (ctl)		16) Motor vehicles & manufactures	veh_man
Animal products nec (oap)		Motor vehicles and parts (mvh)	
Wool, silk-worm cocoons (wol)		Manufactures nec (omf)	
Meat: cattle, sheep, goats, horse (cmt)		17) Services	servs
Meat products nec (omt)		Electricity (ely)	
8) Milk & dairy products	dairy	Gas manufacture, distribution (gdt)	
Raw milk (rmk)		Water (wtr)	
Dairy products (mil)		Construction (cns)	
9) Natural resource extraction	extract	Trade (trd)	
Forestry (frs)		Transport nec (otp)	
Fishing (fsh)		Sea transport (wtp)	
Coal (coa)		Air transport (atp)	
Oil (oil)		Communication (cmn)	
Gas (gas)		Financial services nec (ofi)	
Minerals nec (omn)		Insurance (isr)	
10) Food & beverages	food	Business services nec (obs)	
Food products nec (ofd)		Recreation and other services (ros)	
Beverages and tobacco products (b_t)		PubAdmin/defence/health/educat(osg)	
11) Textiles & leather	tex_lea	Dwellings (dwe)	
Textiles (tex)			

Part II

Trade and the Multifunctionality of Agriculture in EU Countries

6. International trade, agricultural policy reform and the multifunctionality of EU agriculture
A framework for analysis

Janet Dwyer and Hervé Guyomard

Introduction

The main objective of this strand of ENARPRI's work is to clarify how far it is possible for agricultural economists to develop robust analyses to reflect the perceived importance of supporting a multifunctional model of agriculture within the enlarged EU in the context of the ongoing round of multilateral agricultural negotiations. Towards that end, it is necessary to revisit issues of multifunctionality and its meaning in a trade perspective, drawing upon theory and political rhetoric (section 6.1). From this overarching analysis, we move on to consider how agricultural trade agreements, and the Doha talks in particular, are likely to affect the multifunctionality of EU agriculture (section 6.2). This question is addressed principally through examining the likely effects of trade agreements on the policy instruments of the common agricultural policy (CAP), including changes to export refunds, import tariffs and guaranteed prices, as well as the scale and nature of domestic support. Analyses are performed for various EU member states, essentially because the available models to date have been developed at this geographical scale (country case studies). The analyses also cover a varying number and range of indicators of multifunctionality.

6.1 Agricultural policies and multifunctionality: A trade perspective

6.1.1 What is multifunctionality?

In a nutshell, the multifunctional character of farming can be defined as follows. Agriculture is an activity that provides a mix of conventional marketable goods such as food and fibre along with a bundle of non-food and non-fibre outputs that are not marketable in conventional terms. These non-food and non-fibre outputs can be either benefits/goods (e.g. biodiversity preservation) or costs/bads (e.g. biodiversity loss). Most of these non-food and non-fibre outputs have positive/negative externalities or public good/bad characteristics. An externality arises when the action of one economic agent influences either the well being of another consumer or the production possibilities of another producer in an indirect way, i.e. in a manner that is not transmitted by market prices. Pure public goods/bads are defined by two features: they are *non-rival* (consumption of the good/bad by one person does not reduce the consumption available to another person) and *non-excludable* (once the good/bad has been provided to one consumer, it is not possible to prevent others from consuming it). All pure public goods/bads are externalities, but not all externalities are public goods/bads. As agricultural producers do not reap all the benefits of the positive spillovers they provide nor support all the costs associated with the negative spillovers they generate, any market-led situation is likely to be characterised by an under-production of positive externalities/public goods and an over-production of negative externalities/public bads, with respect to the socially optimal level of provision. These are the classic cases of 'market failure' that may provide the legitimacy for public intervention to correct them.

The choice of policy instruments that can be used to correct market failures in respect of multifunctional goods and bads needs to consider at least three important factors. First is the efficiency criterion, i.e. how failures can be corrected in the most efficient way with regard to the use and allocation of resources, including administrative resources. Second are the redistribution effects, i.e. how the policy will alter the distribution of resources and incomes among actors and whether this redistribution matches social perceptions of equity. Third are the trade distortion effects, i.e. which policy instruments will have the least impact in terms of distorting trade.

A strict application of welfare economic theory would lead one to address externalities and public goods/bads following the targeting principle, that is, one policy instrument per market failure. This would consist of letting market forces freely determine the levels of trade, consumption and production of private (marketable) goods while simultaneously addressing externalities and public goods/bads provision through targeted policy instruments, decoupled from production of commodity outputs and coupled only to the provision of non-commodity outputs.[1] This simple and intuitive recommendation immediately raises the very difficult and to date only very partially resolved questions of identifying, measuring and valuing the externalities and public goods/bads associated with agricultural production. Because of this, and also because of jointness in production[2] and transaction costs, this recommendation may be, in practice, neither possible nor desirable (optimal).

Identifying multifunctionality: An impossible mission?

Political rhetoric has adopted a number of implicit meanings for the term 'multifunctionality'. In almost all situations, use of this word embraces the full range of environmental spillovers associated with agricultural activity. In some cases, notably for multifunctionality proponents within the World Trade Organisation (WTO), it also includes food security and the viability of rural areas, as well as some elements of social concern in relation to the culture, customs and networks of traditional, agriculturally-based rural areas (see, for example, Burrell, 2002). Multifunctionality opponents recognise that there are external effects or public goods (or both) stemming from food security and rural area viability. They do not accept, however, that these factors are external effects associated with agricultural

---

[1] In passing, one may note that the targeting principle leads one to decouple agricultural income support instruments from production choices and levels – in this case not to correct an exogenous market failure but to achieve an endogenous policy objective aimed at modifying the redistribution of national income in favour of agricultural producers.

[2] Jointness in production arises when commodity and non-commodity outputs are simultaneously (jointly) supplied – as is often the case with by-products of agriculture such as biodiversity and landscapes.

production. In the case of food security, the externality-generating mechanism would lie on the consumption side and agricultural production would only be a substitute for other sources of supply such as imports or stocks. In the same way, although the viability of rural areas can be related to farming, it is not an externality associated with agricultural production, as the externality-generating mechanism is employment (Rude, 2000). What is clear from the previous analysis is that there cannot be an unambiguous resolution to the problem of identification (Bredhal, Lee & Paarlberg, 2003). In any case, both multifunctionality proponents and opponents agree that agriculture generates negative and positive environmental externalities (OECD, 2001). Unfortunately, they generally disagree on the way these environmental externalities should be addressed.

Environmental externalities: From welfare theory to good practice

In this subsection, let us temporarily assume that multifunctionality is limited to the environment and that the environmental externalities associated with agricultural production can be properly and unambiguously identified, measured and valued (clearly, a very heroic assumption). Under these assumptions, what lessons can be drawn from economic theory, more precisely economic welfare theory?[3]

The first-best results are well known. If there is perfect information and competition in markets, free trade is socially optimal provided that corrective policies properly internalise the positive and negative externalities associated with farming. The second-best results are more interesting. If externalities are not adequately addressed, free trade may not produce the most favourable outcome. Moreover, even in this case, trade policies (export subsidies, import tariffs, import quotas, etc.) are unlikely to be optimal instruments to deal with externalities (Paarlberg et al., 2002; Glebe & Latacz-Lohmann, 2003). According to economic welfare theory, these externalities should be addressed through specific policies following the targeting principle. Yet such specific and targeted policies may be very difficult to define and implement in practice.

Because the instruments that could be used to deal with externalities and achieve a given outcome are not unique, additional criteria need to be

---

[3] For further details, see Guyomard & Le Bris (2003).

used to select the most appropriate instrument (or set of instruments) in any particular case (Fullerton, 2001). These additional criteria include administrative efficiency, the ease of monitoring and enforcement, information needs and uncertainty, political feasibility, equity and distributional effects. Also included is the presence of other distortions (imperfections in other markets) and flexibility and dynamic adjustments (i.e. the flexibility of governments to adjust policy rules as information, measurement and valuation improve along with the flexibility of the economy to adjust the production of commodity and non-commodity outputs). Sensitivity to transaction costs is also very important in influencing these considerations.

In the real world, we should perhaps be willing to accept that the targeting principle as a goal is very rarely either appropriate or achievable. A growing body of contemporary evaluation suggests that policy packages are often the most-effective and efficient approach to dealing with production externalities. These seek to achieve a balanced combination of basic 'sticks' (regulations) to set minimum standards and 'carrots' (incentives) to reward provision beyond the baseline, accompanied by promotion/awareness-raising/training, education and extension activities (i.e. information, to minimise transaction costs). Often, as society grows to understand the relationship between production and externalities over time, the level of the baseline shifts.

A broad definition of multifunctionality

In the framework of ENARPRI, we have retained a broad definition of multifunctionality essentially because our main objective was not to define optimal policies for a multifunctional EU agriculture but to analyse to the extent to which agricultural trade agreements and domestic policy reforms induced by these trade agreements could affect all non-commodity outputs provided by agricultural activity.[4] As a result, potential negative side effects of agricultural change upon multifunctionality include not only negative environmental effects but also effects upon employment and

---

[4] It is also because our purpose is essentially positive that we do not discuss in this chapter the two issues of non-commodity output measurement and valuation (on this point, see, for example, Bohman et al., 1999; Rude, 2000; Randall, 2002; Bredhal, Lee & Paarlberg, 2003).

social cohesion, particularly in the context of marginal or depressed rural areas where farming still represents a significant percentage of economic activity and employment. If agriculture intensifies and capitalises in these areas, this can have a negative impact upon employment and community identity. This pattern is common in several EU regions (in the Spanish steppe agriculture, mountain agriculture, etc.): agricultural jobs contract, people move away or join the unemployed and fragile communities disintegrate. Economists and national policy-makers frequently do not see these consequences as negative, but simply as issues of adjustment. By contrast, local policy-makers often perceive such consequences as significant, negative social side effects of policy changes. Of course, the outcome can be the same when agriculture extensifies and decapitalises, leading to land abandonment and decreased agricultural production. In the same way, the positive side effects of agricultural change upon multifunctionality can include positive environmental spillovers along with the increased viability of rural communities and food security. They may further include social elements such as the support of traditional rural customs, strong social capital or unique cultural heritage.

6.1.2   Multifunctionality and multilateral agricultural negotiations

When the Doha round of multilateral negotiations was launched in November 2001, non-trade concerns were specifically recognised and integrated into the agricultural agenda, albeit to a limited extent.[5] Multifunctionality was at the heart of the talks in the very beginning. But as time goes by, a question arises: Is multifunctionality still a relevant question of the day?

Multifunctionality in the Uruguay round

The WTO makes no judgements about countries' domestic agricultural policy objectives on condition that the instruments used to achieve these objectives have no, or at least minimal, trade-distorting effects. From the

---

[5] Non-trade concerns and multifunctionality can be considered as synonymous. The URAA used the term 'non-trade concerns' instead of multifunctionality, but did not provide a clear definition of the concept. This lacuna largely explains why there is still considerable confusion among WTO countries about what is really meant by the terms non-trade concerns and multifunctionality.

Uruguay Round Agreement on Agriculture (URAA) of 1994, there is in effect an explicit recognition that domestic policies do link with international trade. In practice, the URAA classified domestic policies into three coloured boxes according to their perceived distortion effects on production and trade. The green box included domestic farm programmes that were deemed to be minimally trade-distorting and as a result were exempted from reduction commitments and expenditure limits. It was agreed that non-trade concerns should be addressed using green box instruments.[6]

Multifunctionality in the Doha round

The current round of multilateral negotiations was launched in 2001 at the Ministerial Conference in Doha, Qatar. As far as agriculture is concerned, it has adopted a similar negotiation framework (export competition, market access and domestic support) as was adopted under the Uruguay round. From the very beginning of the round, multifunctionality had its proponents and opponents. Proponents argued that production-linked payments are necessary to ensure the sufficient provision of positive externalities and public goods to the extent that these externalities and public goods are joint products of agricultural production. They also desired an extension of the green box, i.e. more flexibility to classify new kinds of aid as green box programmes. By contrast, opponents argued that

---

[6] The URAA distinguished 11 categories of green box policies: 1) general services, 2) public stockholding for food security purposes, 3) domestic food aid, 4) decoupled income support, 5) government financial participation in income insurance and income safety-net programmes, 6) disaster payments, 7) producer retirement schemes, 8) resource retirement schemes, 9) investment aids, 10) environmental payments and 11) regional assistance. Several categories of policies were designed to address non-trade concerns: food security for programmes (category 2 and possibly category 3), environmental externalities (category 10) and the development of agriculture-based rural area programmes (category 11). Category 4 was specifically defined to address the endogenous policy objective of agricultural income support. Programmes under categories 5 and 6 may be interpreted as aimed at addressing market failures associated with risk and uncertainty, more precisely the incompleteness of risk and uncertainty markets. Finally, programmes under categories 7, 8 and 9 were explicitly designed to ease adjustment and adaptation.

there is an insufficient basis for continuing to offer production-linked payments. They invoked the policy-targeting principle and claimed that there is enough flexibility within the URAA green box definition for green box measures to be used to address all legitimate, non-trade agricultural concerns.

Multifunctionality in agricultural trade talks: Still a relevant question?

The EU was the most important WTO member seeking more flexibility in the design of domestic policy relative to what was provided by the provisions of the URAA green box definition.[7] Japan, Norway, South Korea and Switzerland also took this view. But the reform of the CAP in June 2003 led to widespread decoupling with cross-compliance requirements on income support payments. Modulation is also to be applied in order to achieve a (modest to date) switching of funds from pillar I to pillar II measures. As a result, a large part of CAP domestic support may now be viewed as meeting the criteria for eligibility under the URAA green box definition. If this view is accepted in the context of the current round, the question may be posed as to whether the issue of multifunctionality is still relevant to trade discussions and considerations.

Following the Geneva agreement in August 2005, the Hong Kong Ministerial Conference in December 2005 resulted in a framework agreement for Doha. Detailed commitments have yet to be negotiated, however. The agreement covers the three areas of export competition (with a commitment to eliminate export subsidies by 2013), market access and domestic support. In relation to these three areas, given that export competition is now agreed as an important ingredient and EU agricultural domestic support could be viewed as largely in the green box, then market access may appear as the Achilles' heel of the June 2003–April 2004 CAP reforms. In spite of the successive reforms of the CAP in 1992, 1999 and

---

[7] The positions of the various EU member states on the multifunctionality dossier were, and still are, very heterogeneous. France was clearly the best friend of multifunctionality, essentially because such an attitude allowed French agricultural policy-makers and the main farmers' organisation to claim the maintenance of coupled area and animal payments and hence, to reject any further reform of the CAP in the framework of the so-called 'mid-term review' of the Agenda 2000 CAP.

2003, the world prices of many agricultural products remain much lower than domestic prices in the European Union.

Although it is not possible to define precisely what will be included in the final agreement on agriculture from the Doha round, one can reasonably assume that it will have an impact on EU agriculture, essentially through two transmission channels. Third-country imports should increase, owing to their improved access to the EU market and EU exports should decrease because of the suppression of export subsidies. These changes should have a negative impact on EU market prices and the domestic supply of agricultural commodities. But the Doha round agreement could also affect EU agriculture by inducing (speeding up) further reforms of the CAP (for which domestic pressure is also very likely, notably from enlargement and budgetary discipline). At the very least, the phased elimination of export subsidies would appear to make further reforms to the dairy sector inevitable.

These potential outcomes mean that for those who are genuinely interested in maintaining or indeed enhancing the multifunctionality of EU agriculture, the focus of interest in the trade negotiations and their result is in *how the agreement is likely to affect the balance of non-food benefits and costs generated by the farm sector across the EU*. This analysis needs to be made in the context of recent, significant policy changes at the EU level as represented by the 2003–04 CAP reforms.

6.2 Assessing the effects of agricultural policy reforms on the multifunctionality of EU agriculture

In order to analyse the effects of the 2003–04 CAP reforms and possible Doha outcomes on the multifunctionality of EU agriculture, we need to identify indicators and experiments. More specifically,

- For multifunctionality indicators, what range and types of indicators can be used to measure multifunctionality, in all its different interpretations?

- With regard to policy experiments, what scenarios for future domestic policies and prices can simulate the likely outcomes of the Doha talks within the EU?

On this basis it should then be possible to undertake some impact assessment from which we hope to draw a concluding synthesis and discussion of policy implications.

6.2.1 Multifunctionality indicators

From a literature review, in particular OECD (2001), we have identified six classes of potential indicators that are to a large extent already incorporated in domestic policy impact assessments and thus can be used to examine the implications of change upon multifunctionality (see Dwyer et al., 2005, for more details). These six classes correspond to

1) economic indicators (prices, supplies, exports and imports, farm or household incomes, etc.);

2) primary factor use (capital, land and labour);

3) environmental indicators (greenhouse gas, pollution by nitrates or fertilisers as a whole, pollution by pesticides, soil conservation, water management, biodiversity preservation, landscape preservation, etc.);

4) agricultural structure indicators (number of farmers and farm workers, farm sizes and types, etc.);

5) farm management indicators (set-aside areas, land abandonment, livestock density, etc.); and

6) cultural and social indicators (agriculturally linked customs and events, percentage of rural population connected with farming and the proportion of locally sourced food sold in rural areas).

There are significant measurement problems for many of these indicators, as well as potentially ambiguous interpretations for some of them depending upon how one values certain issues (such declining farm employment, which can be seen as a positive efficiency adjustment or a negative issue, notably in depressed rural areas). There is also unequal coverage of the six classes of indicators by the existing range of available quantitative models. Aspects associated with cultural heritage and social capital/values are not included in any of the models and thus can only be estimated by *ex-post* qualitative (expert) assessments. In addition, domestic policies vary among countries owing to the increased subsidiarity in CAP decision-making, which now affects the scale and pattern of support under both pillars. These issues give rise to a problem in the extrapolation of results obtained for one country to any other member states.

A final remark is also needed. It is clear that the time horizon of simulations matters because environmental and social effects arising from agricultural change frequently take some time to become fully apparent. Thus in general, our country case studies have been undertaken with a medium-term horizon of 2013–15 in mind.

6.2.2 Policy experiments

The baseline: The June 2003–April 2004 CAP reform (S1)

The June 2003–April 2004 CAP reform scenario is the baseline against which effects arising from the Doha round of negotiations can be assessed. It must be remembered that this scenario gives rise to different implementation rules in each member state. Such rules include single farm payments on a historical or regional basis, partial recoupling of some pillar I direct aids or the use of national envelopes in a number of member states, different requirements as regards cross-compliance and good agricultural and environmental practices, and the varying balance, focus and scale of pillar II schemes. For any single country where models exist, it is possible to examine the impact of the June 2003–April 2004 CAP reform on market prices, product supply and factor demand, intensification, land use, land abandonment, farm income and more rarely the number and the size of farms. Yet given that most models have primarily been developed to examine the effects of changes in pillar I market measures on the agricultural sector, there are important limits to the degree of policy detail that the models can incorporate. Cross-compliance and pillar II measures, as well as entry and exit from agriculture, are frequently not modelled.

A fully decoupled June 2003–April 2004 CAP reform (S1bis)

To draw a contrast with this baseline, it is desirable to consider a fully decoupled June 2003–April 2004 CAP reform with no partial recoupling and common cross-compliance requirements. By comparing this alternative implementation of the June 2003–April 2004 CAP reform with the baseline, it should be possible (at least in theory) to assess the impact of the partial recoupling of some pillar I direct aids – an option taken up by a number of member states.

In the scenarios for both the baseline and the fully decoupled June 2003–April 2004 CAP reform, trade instruments (export subsidies and import tariffs) are assumed to be unchanged at base period levels.

A fully decoupled CAP – Decoupling extended to the dairy sector, export subsidy abolition and improved access to the EU market (S2)

The third policy experiment corresponds to a fully decoupled CAP extended to the dairy sector. Export subsidies are abolished and market access is gradually increased to 10% of EU domestic consumption. This scenario represents the impact of a fully decoupled CAP in the context of a Doha agreement entailing export subsidy suppression, assuming the URAA definition of coloured boxes is retained (meaning that there is no problem for the EU to respect its commitments on the domestic support dossier) and EU imports from third countries increases.

A fully decoupled CAP with reduced, decoupled direct aids (S2bis)

In an attempt to capture the possibility that either as a result of Doha round or talks beyond it the single farm payment is not accepted as a green box aid, it is interesting to examine a version of scenario S2 such that decoupled direct aids are reduced by 20%. This scenario assumes that the EU is forced to reduce decoupled direct aids because of WTO commitments, domestic pressure or notably budget constraints (enlargement to Bulgaria and Romania, with decoupling extended to nearly all supported sectors).

A fully decoupled CAP with resources shifted from pillar I to pillar II (S3)

If the driving force for the scenario S2bis change was the WTO, one could also consider a modification that would not reduce the overall support under the CAP but shift it between the pillars, i.e. a fully decoupled CAP with an increased share of resources devoted to pillar II measures. More specifically, this scenario S3 assumes that funds saved through a 20% decrease in pillar I measures are transferred to pillar II, in order to achieve environmental and social objectives.

This range of scenarios will be quite helpful in assessing the potential effects of a Doha agreement on EU agriculture in different member states, even if not all the scenarios can be examined by all the models. In tracking Doha progress to date, the existing level of decoupling seems assured while increased market access will probably compel further reforms and thus probably more decoupling. Export subsidies will be phased out. Yet while there will not be a lot more EU funding for pillar II (following the last agreement on the EU financial perspectives), there could be some interest among member states in considering the scope for mitigating the effects of reform by using modulated money to target particular measures under pillar II, to increase the goods and minimise the bads.

6.3   The models for our study

Most of the existing models used for assessing the impact of multilateral agricultural negotiations focus on prices, incomes, production and environmental outcomes. The multifunctional aspects of policy reforms are usually reduced to a relatively narrow set of indicators, mostly linked to environmental issues, but some social aspects, notably farm employment, can also be considered. Table 6.1 gives a summary of the characteristics of the models used for this ENARPRI examination of multifunctionality.

Table 6.1 Main characteristics of the models used for examining multifunctionality

Country, model name and type	Scenarios modelled	Indicators of multifunctionality
Ireland FAPRI-Ireland is a set of econometric, dynamic, multi-product, partial equilibrium commodity models	CAP reform 2003 with full decoupling as the baseline, compared with a Doha scenario of increased market access (60% average tariff cut), 70% cuts in the amber- box domestic support (aggregated measurement of support) and elimination of export subsidies over 10 years	Greenhouse gas emissions, ammonia emissions, fertiliser use and farm incomes
Finland DREMFIA is a partial equilibrium recursive model with 17 production regions, combining technology diffusion with optimising producer and consumer returns	CAP reform 2003 with partial decoupling; full decoupling across the EU; new environmental cross-compliance to create uncropped field margins on all arable land; pillar I cuts by 20% by 2013.	Production levels, agricultural land use, nutrient balances, biodiversity, pesticide usage, farm income and farm labour
Greece National and regional social accounting matrices	All specified scenarios	Production levels, farm/other employment, agricultural land use and pollution emissions
Czech Republic FARMA–4 (linear programmes optimising production for given outcomes based on 'typical' farms)	Pre-CAP policy as in 2002, compared with full decoupling or 100% uptake of an agri-environment scheme	Farm labour, farm types/systems, livestock numbers, grassland cover, fertiliser and energy use

Source: Authors' compilation.

In Finland, the team used the *DREMFIA model,* a dynamic, regional sector model of Finnish agriculture, in which a technology diffusion model is combined with an optimisation routine that stimulates annual production decisions (within the limits of fixed factors) and price changes. The scenarios examined included the actual CAP reform of 2003 as implemented in Finland with some sectors still partially coupled, along with full decoupling and the application of a simplified cross-compliance regime for compulsory arable field margins, as well as a scenario involving 20% cuts in CAP pillar I payments. The list of indicators applied in the model is longer than in any of the other models described or used by other ENARPRI partners (see Table 6.1). It includes production levels, land use, nutrient balances, biodiversity index, pesticide use, farm incomes and employment; however, the model could not deal with a scenario involving the expansion of pillar II funding.

In Ireland, the team used a *FAPRI-Ireland model,* or a set of dynamic, multi-product, partial equilibrium commodity models, built on a similar platform as the FAPRI models developed in the US. The Irish researchers examined the scenario of the CAP reform in 2003 implemented as full decoupling, versus an expected Doha outcome involving more market access and the phasing out of export subsidies. The indicators examined for multifunctionality included greenhouse gas emissions, forestry carbon sequestration, ammonia emissions and fertiliser usage. FAPRI is capable of dealing directly with the outcomes of international trade agreements rather than having to transform these into domestic policy changes. Yet the Irish model is less readily able to deal with the implications of some of the domestic policy scenarios outlined in this chapter, especially where they involve making assumptions about CAP pillar II spending and its effects.

In Greece, the team investigated changes using national and regional *social accounting matrices (SAMs)* constructed to examine changes in the farm and non-farm sectors at these levels in response to policy changes. They also examined all four specified scenarios for this project, because the approach included consideration of the effects of CAP pillar II policies (in Greece, notably more than 70% of pillar II spending is allocated to improving farm competitiveness (European Commission, 2003)). The indicators analysed were farm output, farm/other employment, agricultural land use and levels of pollution emissions.

In the Czech Republic, the team used a model called *FARMA–4,* which is a non-linear optimising model based on the optimising behaviour of three main farm types, linked to certain multifunctionality indicators. The scenarios examined were the CAP policy as in 2002, full decoupling and the expansion of a simple grassland-conserving agri-environment scheme across the country. The indicators tracked were employment, production types, stock numbers, grassland cover, fertiliser and energy use.

In summary, the various models and approaches applied by the ENARPRI team can cover all the scenarios. The biggest modelling difficulties clearly apply to scenario S3, because few models have adequate methods to represent the range of pillar II measures deployed. Wherever the models have difficulties, additional ad-hoc information and qualitative/policy evaluative expertise can be used to examine the implications for multifunctionality. Thus it is indeed possible to seek to identify the links between trade policies and multifunctionality, albeit only at the level of individual EU member states.

Concluding remarks

This chapter has sought to provide the context for the work of the ENARPRI team on the relationships between multilateral trade agreements and multifunctionality – specifically, examining implications for the multifunctionality of agriculture in the EU. Framed in this context, the country case studies contribute a wealth of data and potential insights into these implications and their likely outcomes in terms of a range of environmental, social and economic indicators. A final section in this strand of ENARPRI work is presented in chapter 11, which synthesises these findings and analyses their relevance both to the understanding of trade and multifunctionality interrelations, and to the domestic and international policy process.

References

Bohman, M., J. Cooper, D. Mullarkey, M.-A. Normille, D. Skully, S. Vogel and E. Young (1999), *The use and abuse of multifunctionality*, US Department of Agriculture, Economic Research Service, Washington, D.C.

Bredhal, M., J.-G. Lee and P.-L. Paarlberg (2003), *Implementing Multifunctionality*, paper presented at the "International Conference on Agricultural Policy Reform and the WTO: Where are we heading?" held in Capri on 23–26 June.

Burrell, A. (2002), *Multifunctionality, Non-Trade Concerns and the Millennium Round*, paper presented at the Société Française d'Economie Rurale (SFER) "Conference on Multifunctionality", held in Paris on 22 March.

Dwyer, J., D. Baldock, H. Guyomard, J. Wilkin and D. Klepacka (2005), *Scenarios for Modelling Trade Policy Effects on the Multifunctionality of European Agriculture*, ENARPRI Working Paper No. 10, CEPS, Brussels, January.

European Commission (2003), *Factsheet: Overview of the implementation of rural development policy 2000-2006 - Some facts and figures*, European Commission, Brussels.

Fullerton, D. (2001), "A Framework to Compare Environmental Policies", *Southern Economic Journal*, Vol. 68, No. 2, pp. 224-48.

Glebe, T., and U. Latacz-Lohmann (2003), *Multifunctionality and free trade: A formal analysis*, paper presented at the "Agricultural Economics Society (AES) Conference", held in Seale Hayne, UK on 11–14 April.

Guyomard, H. and K. Le Bris (2003), *Multifunctionality, Agricultural Trade and WTO Negotiations: A Review of Interactions and Issues*, ENARPRI Working Paper No. 4, CEPS, Brussels, December.

OECD (2001), "Executive Summary", in *Multifunctionality: Towards an analytical framework*, OECD, Paris.

Paarlberg, P.-L., M. Bredhal and J.-G. Lee (2002), "Multifunctionality and agricultural trade negotiations", *Review of Agricultural Economics*, Vol. 24, No. 2, pp. 322-35.

Randall, A. (2002), "Valuing the outputs of multifunctional agriculture", *European Review of Agricultural Economics*, Vol. 29, No. 3, pp. 289-307.

Rude, J. (2000), *Appropriate remedies for non-trade concerns*, CATRN Paper, Canadian Agri-Food Trade Research Network.

7. Evaluating the impact of alternative agricultural policy scenarios on multifunctionality
A case study of Finland

Heikki Lehtonen, Jussi Lankoski and Jyrki Niemi

Introduction

Evaluating changes in such a broad and often vaguely defined area as multifunctionality is by no means an easy task. It may be that the vagueness of concepts such as 'sustainable development' and 'multifunctionality', which can be understood in many ways depending on the context, is in fact what makes them such popular policy goals. In this chapter we try to evaluate the effects of agricultural policies on different dimensions of multifunctionality in a consistent manner. Since we are agricultural economists, our tool is microeconomics-based economic modelling. This approach may be considered biased since certain dimensions of multifunctionality refer to biological and social relationships that may not be sufficiently taken into account in economic models. Nevertheless, in economic logic we find an approach that provides a sufficiently simple way of reasoning that facilitates a consistent analysis. Particular emphasis should be given to its different dimensions. For example, biological relationships, physical material flows and social phenomena require special efforts when analysing the effects of very different agricultural policies on some selected multifunctionality indicators, and we feel that a simple calculation of many indicators on the basis of existing economic models may provide misleading results. A specific research project may be required to clarify and set up the relevant production functions and relationships necessary for a credible assessment of the indicators of multifunctionality.

The rest of this chapter is organised as follows. In section 7.1 we clarify our concept of multifunctionality, as well as our principal approach and the main challenges to evaluating it. Section 7.2 presents the agricultural sector model employed by the study as well as the selected multifunctionality indicators used in the calculation. In section 7.3, alternative agricultural policy scenarios are listed and interpreted. Section 7.4 lays out the indicator results from the sector model. Finally, the chapter concludes with an evaluation of the results and a discussion of the applicability of the chosen approach.

7.1 Definition of multifunctionality and our approach to its evaluation

The notion of *multifunctional agriculture* refers to the fact that agricultural production provides not only food and fibre but also different non-market commodities. These non-commodity outputs include the impact of agriculture on environmental quality, such as rural landscape, biodiversity and water quality. Often this list also includes the socio-economic viability of the countryside, food safety, national food security and animal welfare together with cultural and historical heritage. There is no universally accepted definition of multifunctionality, and the emphasis given to various types of non-commodities differs. OECD (2001) provides a working definition of multifunctionality. This definition sets out the fundamentals of multifunctionality as: i) the existence of joint production of commodity and non-commodity outputs and ii) the characteristic that some non-commodity outputs exhibit qualities of externalities or public goods (OECD, 2001, p. 13). Both theoretical and applied work has tried to push forward this working definition.

Academic research on multifunctionality has mainly focused on the environmental dimension of multifunctionality. The reason for this is evident: Pareto optimality requires that all positive and negative externalities should be internalised, thus giving a firm theoretical basis to the environmental dimension multifunctionality. Boisvert (2001), Romstad et al. (2000), Guyomard & Levert (2001), Anderson (2002), Paarlberg et al. (2002), Vatn (2002), Peterson et al. (2002), Lankoski & Ollikainen (2003), Guyomard et al. (2004) and Lankoski et al. (2004) focus on the properties and policy design of multifunctional agriculture in either a closed economy or an international trade framework. All these studies approach multifunctionality with the help of the theory of joint production.

None of the previous papers has focused on the non-public good aspects (such as rural viability or food security) of multifunctional agriculture. The decision of whether features other than these public goods should be introduced to the social welfare function of agriculture is a complex question. As observed by OECD (2001) and argued by Anderson (2002) for instance, food security and rural viability cannot entirely be subsumed into the category of public goods. Very recently Ollikainen & Lankoski (2004) have enlarged the conventional public goods and bads framework to include non-public good qualities, more specifically that of rural viability. Following OECD (2001), they express rural viability through the employment effects of agricultural production. They demonstrate that rural viability moderates policy towards public goods (environmental non-commodity outputs) because society trades-off public good qualities for those associated with rural viability.

An important and so far nebulous issue in the literature on multifunctional agriculture is the fact that joint production of commodity and non-commodity outputs naturally differs among alternative production lines. For example, milk production can be considered as a truly multifunctional production activity (in terms of both environmental non-commodity outputs and rural viability aspects), which nevertheless differs considerably from the multifunctionality associated with crop production. Previous literature has focused on policy packages mainly related to crop production (see e.g. Guyomard et al., 2004; Lankoski et al., 2004), but not on alternative production lines. Hence, an important research question arises: How does multifunctional agricultural policy affect the relative profitability, public goods and viability aspects across alternative production lines within agriculture? This chapter considers this question, focusing on relative profitability and changes in the supply of joint non-commodity outputs as a result of changes in profitability. Moreover, we analyse the merits of environmental cross-compliance schemes to address multifunctionality.

We examine this problem through a dynamic, regional sector model of Finnish agriculture (DREMFIA) (for a thorough description of the model, see Lehtonen, 2001). This model is employed to assess the effects of alternative policy scenarios on the multifunctional role of Finnish agriculture. In terms of environmental non-commodity outputs, we focus on nutrient runoffs, landscape diversity and biodiversity. We use regional nutrient surpluses (soil-surface balance method) for nitrogen and phosphorus as a proxy for nutrient runoffs and resulting surface-water quality impairment. Physical input flows are particularly relevant in analysing environmental effects and adjustments to agricultural policies. Evaluation of nutrient balances in very different agricultural scenarios is a challenging task since drastic changes in policies may imply larger changes in the use of fertilisers, manure and feedstuffs and in the resulting crop and animal yields than historically observed. First, it is crucial that total *quantities* of inputs are validated to observed aggregate levels in the economic model and not just the total *values* of inputs and outputs. Second, more data should be used besides historical farm-level data on physical inputs and outputs. Experimental and research data may reveal important relationships between physical inputs and outputs. The available relevant crop and animal feeding research results have been used in setting the production functions in the model, while the level of physical inputs and outputs is validated to observed levels. These two aspects are not always easy to combine. Nevertheless, if the model is also consistent in terms of the physical flows of inputs and products, and in terms of yield responses to large changes in prices and supports, then the indicator changes may be credible as well, up to our current knowledge of the biological production process. The continuous updating of farm-level and biological research information is necessary. The calculation of many agri-environmental indicators should not be done as a careless routine. On the contrary, it requires conscientious assessment: Does the data used in the economic model tell us what will happen in the production process in very different policy scenarios?

Once the crucial relationships of the bio-physical production process have been determined using the best available data and knowledge, the resulting changes in production facilitate a rather straightforward calculation of a number of indicators. For assessing landscape diversity, we employ Shannon's diversity index (SHDI) to assess the richness (number of different land-cover classes, i.e. cultivated crops and bare and green fallow) and evenness (uniformity of distribution for types of land use) of agricultural land use under different policy scenarios. The area designated as providing different types of wildlife habitats is used as a proxy for biodiversity.

As regards other non-commodity outputs, one of our interests is rural socio-economic viability. In line with Ollikainen & Lankoski (2004), we describe the core economic content of rural viability by employment in agriculture and in the rural sectors serving agriculture. Our approach is to use an economic sector-level model that does not explicitly consider linkages between agriculture and other sectors of the national economy. Hence the approach cannot provide any final results concerning the employment effects of agricultural policies. The approach taken here, however, does consider in a detailed way the linkages among different production lines in agriculture. Especially the changes in the relative profitability of various agricultural activities affect production volumes and land use since land is a restricted, limited resource for agricultural activities. Agricultural land and production equipment (buildings and machinery), on the other hand, are of low value in alternative uses in Finland where the countryside is relatively sparsely populated. This characteristic makes even a partial equilibrium approach interesting in evaluating the socio-economic effects of agricultural policies.

7.2 The model and multifunctionality indicators

7.2.1 The sector model

DREMFIA is a dynamic recursive model that includes 17 production regions. The model provides effects of various agricultural policies on land use, animal production, farm investments and farmers' income. The model consists of two major parts: 1) a technology diffusion model that determines sector-level investments in different production technologies; and 2) an optimisation routine that simulates annual production decisions (within the limits of fixed factors) and price changes, i.e. supply and demand reactions, by maximising producer and consumer surpluses subject to the balance and resource (land and capital) constraints of regional products (Figure 7.1).

In the DREMFIA model, annual land use and production decisions from 1995 to 2020 are simulated by an optimisation model that maximises producer and consumer surplus subject to regional product balance and resource (land) constraints. Final products and intermediate products may be transported between the regions. The optimisation model is a typical, spatial, price equilibrium model (see for example Cox & Chavas, 2001), except that no explicit supply functions are specified (i.e. supply is a primal specification). Furthermore, foreign trade activities are included in DREMFIA. The Armington assumption (Armington, 1969), which is a common feature in international agricultural trade models but less common in one-country sector models, is used. Imported and domestic products are imperfect substitutes, i.e. endogenous prices of domestic and imported products are dependent. There are 18 different processed milk products and their associated regional processing activities in the model.

Figure 7.1 Basic structure of the DREMFIA mode

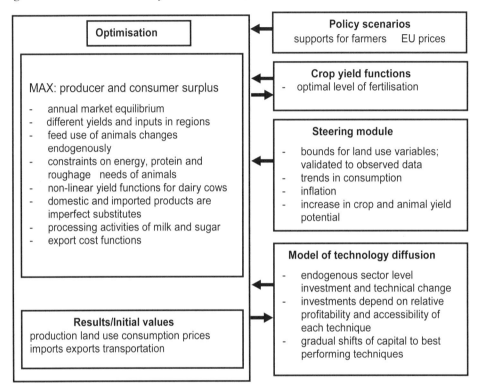

Four main areas are included in the model: southern Finland, central Finland, Ostrobothnia (the western part of Finland) and northern Finland (Figure 7.2). Production in these areas is further divided into sub-regions on the basis of the support areas. In total, there are 17 different production regions. This allows a regionally disaggregated description of policy measures and production technology. The final and intermediate products move between the main areas at certain transportation costs.

Technical change and investments, which imply the evolution of farm-size distribution, are modelled as a process of technology diffusion. Investments are dependent on economic conditions such as interest rates, prices, support, production quotas and other policy measures and regulations imposed on farmers. The model of technology diffusion used generally follows the main outlines of Soete & Turner (1984).

Figure 7.2 Main regions in the dynamic regional sector model of Finnish agriculture (DREMFIA)

Two crucial aspects about diffusion and adaptation behaviour are included: first, the profitability of a new technique; and second, the risk and uncertainty involved in adopting a new technique. The information about and likelihood of adopting a new technique will increase as its use becomes widespread.

To cover the first aspect, the likelihood of adopting a new technique ($f_{\beta\alpha}$) is made proportional to the fractional rate of profit increase in moving from technique α to technique β, i.e. $f_{\beta\alpha}$ is proportional to $(r_\beta - r_\alpha)/r_\alpha$, where r_α is the rate of return for technique α and r_β is the rate of return for technique β. The second aspect is modelled by letting $f_{\beta\alpha}$ be proportional to the ratio of the capital stock in β technique (K_β) to the total capital stock K (in a certain agricultural production line), i.e. K_β/K. The total investments to α technique, after simplification, is

$$I_\alpha = \sigma(Q_\alpha - wL_\alpha) + \eta(r_\alpha - r)K_\alpha \tag{1}$$

where σ is the savings rate (proportion of economic surplus re-invested in agriculture), η is the farmers' propensity to invest in alternative techniques, Q_α is the total production-linked revenue for technique α, w is a vector of input prices, L_α is a vector of variable production factors of technique α, and r is the average rate of return on all techniques.

The interpretation of the investment function is as follows. If the value of η was zero, then (1) would show that the investment in α technique would come entirely from the investable surplus generated by α technique. For $\eta \neq 0$, the investment in α technique will be greater or less than the first term on the right-hand side, depending on whether the rate of return on α technique is greater or less than the average rate of return on all techniques (r). This seems reasonable. If a technique is highly profitable it will tend to attract investments, and conversely, if it is relatively less profitable investments will decline. If there are no investments in α technique at some time period, the capital stock K_α decreases at the depreciation rate. To summarise, the investment function (1) is an attempt to model the behaviour of farmers whose motivation to invest is greater profitability, but who, nevertheless, will not adopt the most profitable technique immediately because of uncertainty and other constraining factors.

The investment function (1) shows that the investment level is strongly dependent on the capital already invested in each technique. This assumption is consistent with the conclusions of Rantamäki-Lahtinen et al. (2002) and Heikkilä et al. (2004), i.e. that farm investments are strongly correlated with earlier investments, but poorly correlated with many other factors, such as liquidity or financial costs. Other common features, except for the level of previous investments by investing farms, were hard to find. Hence, the assumption made on cumulative gains from earlier investments seems to be supported by the findings of Rantamäki-Lahtinen et al. (2002) and Heikkilä et al. (2004). The investment function allows regional re-location and concentration of production and technical change at the same time.

Three dairy techniques (representing α techniques) and corresponding farm-size classes have been included in the DREMFIA model: farms with 1-19 cows (labour-intensive production), farms with 20-49 cows (semi-labour intensive production) and farms with 50 cows or more (capital-intensive production). The parameter σ has been fixed to 1.07, which means that the initial value 0.85 (i.e. farmers re-invest 85% of the economic surplus on fixed factors back into agriculture) has been scaled up by 26%, which is the average rate of investment support for dairy farms in Finland. The value of η (fixed to 0.77) is then used as a calibration parameter, which results in investments that facilitate the *ex post* development of dairy farm structure and milk production volume. The chosen combination of the parameters σ and η (1.07:0.77) is unique in the sense that it calibrates the farm-size distribution to the observed farm-size structure in 2002 (Farm Register, 2002). Choosing larger σ and smaller η exaggerates the investments on small farms, and choosing smaller σ and larger η exaggerates the investments on large farms. Choosing smaller values for both σ and η results in investment and production levels that are too low, and choosing larger values for both σ and η results in overestimated investment and production levels compared with the *ex post* period.

Use of variable inputs, such as fertilisers and feedstuffs, is dependent on agricultural product prices and fertiliser prices through production functions. The nutrients from animal manure are explicitly taken into account in the economic model. The feeding of animals may change provided that nutrition requirements, such as energy, protein, phosphorous and roughage needs, are fulfilled. In the feasible range of inputs per animal, production functions can be used to model the dependency between the average milk yield of dairy cows and the amount of concentrates and other grain-based feedstuffs. Since in historical farm-level data there are relatively fewer instances of low or high levels of concentrates, the dataset is enriched by experimental data. A number of trials have been undertaken by agro-biological research on the yield-response effects of significant changes in animal feeding and crop fertilisation (Sairanen et al., 1999 and 2003; Bäckman et al., 1997). In the case of dairy cows and field crops, the uniform pattern of the results of many similar trials facilitates the inclusion of the data material in the estimation of the production functions. Hence, the production functions in the model include not only the observed historical variation in the use of inputs but also responses to large changes in the use of inputs rarely observed in actual farms. This approach means that if agricultural policies imply significant changes in feed or fertiliser use, for example through relative input and output prices or restrictions on land use, the most relevant biological relationships are taken into account in calculating the economically-rational production adjustment. The new farm-level data on animal feeding, however, is richer (there is more variation) than previous data as farmers have gradually increased the use of feed concentrates (Pro Agria, 2005), so we believe that the role of experimental data decreases over time.

Milk quotas, which constrain milk production at the farm and country level, are traded within three separate areas in the model. Within each quota-trade area, the sum of quotas purchased must equal the sum of quotas sold. The price of the quota is the weighted sum of the shadow values of an explicit quota constraint in each sub-region. Milk quota trade has an important role in facilitating improvements in production efficiency. The observed milk quota prices have served as a valuable reference point in the model validation.

The overall model replicates very closely the *ex post* production development in 1995-2003. Official agricultural production and price statistics[1] have been used as the basis in validation. Calibrating the unobserved parameters of the investment model (discussed above) is a significant part of the overall validation of the model. Price changes in 1995-2003 have been validated through calibrating the unobserved parameters in the Armington system and in export-cost specification (see Lehtonen, 2001, for details). The total value of each single input (calculated from the input specifications of many production activities in the model) has been checked and validated by using cross-sectional statistical data (Statistics Finland, 1995 and 2003). Furthermore, total *quantities* of inputs and not just the total *values* of inputs and outputs are validated to observed aggregate levels. Hence, the validation of the model is also consistent in terms of the physical flows of inputs, such as fertilisers and feedstuffs. Physical input flows are particularly relevant in analysing environmental effects and adjustments to large changes in agricultural policy. In the validation process, two individual years were excluded due to very unusual weather conditions and subsequent crop failures.

The long- and medium-term changes in aggregate amounts and regional location of production are consistent in the economic sense as the model is built to reach a steady state of equilibrium in a 10-15 year period given no further policy changes. There is a built-in gradual adjustment in

---

[1] See the website http://matilda.mmm.fi.

the model as fixed production factors and animal biology make immediate adjustments costly. Non-linear production functions in the model are concave, i.e. the marginal productivity decreases with output. The steady-state equilibria found at the whole country level are also the result of the limited domestic consumption of foodstuffs and expensive exports, owing to low EU prices against the production and transportation costs. Another reason for steady states in a 10-15 year period is the Armington assumption and the expectation that consumers have some preferences as to domestic products, i.e. the scarcity of domestic foodstuffs slightly increases producer prices, even though this increase is relatively small (only 1-10% on the producer price level) in the model, when validated to the observed price development. A more detailed presentation of the model and its parameters can be found in Lehtonen (2001, partially updated in 2004).

7.2.2 Multifunctionality indicators

The available indicators derived from the DREMFIA model output are listed in Table 7.1. Not all the indicators are listed in this chapter, however. Most of the indicators presented in this study, such as production volumes, hectares of crops, nutrient balances and incomes, as well as direct agricultural employment, are calculated directly using the DREMFIA sector model. Indirect agricultural employment was calculated using regional input–output tables that take into account both upstream and downstream indirect employment (Knuuttila, 2004).

Table 7.1 The applied indicators, derived from the DREMFIA model, in the agricultural policy scenario analysis

Applied indicator	Measured quantity	Indicator reflecting	Strategic goal of the indicator
Total number of animal units up to 2020	Animal units	The scale and long-term economic viability of aggregate animal production	To ascertain the relative economic viability of animal production in different policy scenarios
Number of bovine animal units	Animal units	The scale and long-term economic viability of dairy and beef production	To ascertain the relative economic viability of dairy and beef production in different policy scenarios

Table 7.1 cont.

Number of pig animal units	Animal units	The scale and long-term economic viability of pig production	To ascertain the relative economic viability of pig production in different policy scenarios
Number of poultry animal units	Animal units	The scale and long-term economic viability of poultry production	To ascertain the relative economic viability of poultry production in different policy scenarios
Total cultivated area (excluding set-aside) up to 2020	Hectares	Incentives for active crop production	Changes in incentives for active crop production
Set-aside area	Hectares	Incentives for fulfilling cross-compliance criteria and minimising costs	Changes in incentives in fulfilling cross-compliance criteria and minimising costs in different policy scenarios
Unused area	Hectares	Share of abandoned agricultural land owing to unprofitable production	Changes in the share of abandoned land owing to unprofitable production
Grass area	Hectares	The scale of grass feed production; incentives for grass feed use and bovine animal production	Changes in the scale and incentives for grass feed production in different agricultural policy scenarios
Grain area	Hectares	The scale and incentives for grain production	Changes in the scale and incentives for grain production in different policy scenarios

Table 7.1 cont.

Nitrogen balance on cultivated area[a]	Kg/ha	Nitrogen-leaching potential from cultivated land	Changes in the nitrogen-leaching potential in different policy scenarios
Phosphorous balance on cultivated area[a]	Kg/ha	Phosphorous-leaching potential from cultivated land	Changes in phosphorous-leaching potential in different policy scenarios
Habitat index	(Scale of 0-100)	Value of agricultural land for certain indicator species	To ascertain the biodiversity effects of agricultural policies
Agricultural income	€ million	The level of economic activities in agriculture	Changes in the level of economic activity in different policy scenarios
Profitability coefficient[b]	–	Profitability of agricultural production	Changes in profitability of agricultural production in different policy scenarios
Labour hours in agriculture; indirect effects on employment	Million hours or 1,000 employees	Social sustainability of farmers, the working conditions of agricultural labour	Changes in the number of people employed in agriculture and related professions in different policy scenarios
Agricultural income per hour of labour	€/hour	Economic and social welfare of farmers	Changes in the economic and social viability of agriculture in different policy scenarios

[a] The soil surface nitrogen and phosphorus balances are calculated as the difference between the total quantity of nitrogen or phosphorus inputs entering the soil and the quantity of nitrogen or phosphorus outputs leaving the soil annually, based on the nitrogen or phosphorus cycle.

[b] The profitability coefficient is a ratio obtained when the agricultural surplus is divided by the sum of the entrepreneur family's salary requirement and the interest requirement on the capital invested.

Source: Authors' compilation.

Shannon's diversity index (*SHDI*) was applied in the land-cover diversity calculations (McGarigal & Marks, 1995). The index is based on information theory (Shannon, 1948) and it is frequently used in diversity quantifications (Di Falco & Perrings, 2003; Hietala-Koivu et al., 2004). The values of *SHDI* were calculated according to the formula:

$$SHDI = - \sum_{i=1}^{m} \left(P_i \times \ln P_i \right), \qquad (2)$$

where *m* is the number of land-cover classes, P_i measures the proportion of area covered by land-cover type *i* and ln denotes natural logarithm. *SHDI* is equal to zero when the agricultural area contains only one land-cover class (i.e. no diversity). The value of Shannon's diversity index increases as the number of different land-cover classes increases or the proportional distribution of the area among land-cover classes becomes more equitable (or both). Hence, for a given number of land-cover classes, *SHDI* reaches its maximum when the proportions of land-cover classes are uniform, i.e. $P_1 = P_2 = \ldots = P_m = 1/m$ (McGarigal & Marks, 1995).

In addition to diversity in agricultural land use, we also consider the potential biodiversity effects of policy scenarios. According to Duelli (1997), biodiversity evaluation at the regional level can be based on landscape parameters. Even though landscape diversity indicators give an overview of biological diversity, there are no general models that relate overall species diversity to landscape diversity (Jeanneret et al., 2003). The relationship thus depends strongly on the organism examined.

The aggregate soil surface balances (surplus/deficit) for nitrogen and phosphorus per cultivated area, excluding set-aside, were calculated by adding the nutrient content of fertilisers, organic manure and nitrogen deposits, and by subtracting the nutrient content of the harvest and losses to the atmosphere. The calculated net nutrient surplus (kg/ha) provides an indicator of the production intensity and of the potential nutrient losses and environmental damage to surface and ground waters.

The size of the pesticide application area was also reported. Chemical pesticides enhance agricultural productivity but also pose potential risks to human health and the environment. For example, they may cause the contamination of surface water.

The habitat index was calculated on the basis of a large-scale dataset of empirical observations concerning butterfly numbers on lands farmed by different crops (Kuussaari & Heliölä, 2004). The butterfly was selected as an indicator species by environmental scientists. It was observed that green set-aside was a more valuable habitat for butterflies by six times than grain fields, whereas field edges provided more valuable habitats by more than seven times and natural meadows by more than ten times. These relative weights were used directly when calculating a habitat index as a linear vector, divided by overall hectares of agricultural lands. Hence, the resulting index represents an average biodiversity value of all agricultural land in comparison with natural meadows. In 1995-2004 the calculated habitat index was valued at between 20 and 25. The index would be 100 if all agricultural lands were changed to natural meadows.

7.3   Alternative agricultural policy scenarios

Based on the current multilateral trade negotiations and the most recent indications of negotiation positions from various World Trade Organisation (WTO) members, there are a number of interesting policy scenarios that could be analysed. In response to the WTO's framework agreement approved in Geneva on July 2004 and the possible conclusion of the Doha round, the following scenarios are formulated as below.

7.3.1   The baseline scenario (BASE)

The baseline scenario (BASE) corresponds to the continuation of the Agenda 2000 agricultural policy (agreed in Berlin in 1999) over the medium term. The purpose of the baseline is not to forecast the future but to establish a yardstick against which policy simulations can be judged. The baseline simulation is a view of the world in which policies remain unchanged. The impact of EU enlargement has not been incorporated into the baseline. It is important, however, to remember that the baseline scenario includes the reductions in intervention support prices and future increases in quotas in the dairy sector that were politically agreed in Berlin in 1999. Therefore, it is assumed that the producer price of milk in the EU and in Finland would fall by 12% from the average producer price of 1999-2001 by 2008.

7.3.2   Common agricultural policy reform scenario (REF)

The ongoing common agricultural policy (CAP) reform scenario (hereafter the 'REF' scenario) follows the CAP reform agreement made in June 2003, according to which most direct CAP subsidies will be decoupled from production and paid in a single, lump-sum farm payment based on 2000-02 historical production levels (European Commission, 2003). On options given for the EU member states, the Finnish government has decided that the implementation of the reform will start in 2006. From there on, all CAP arable-area payments will be decoupled from production and a regionalised flat-rate payment will be paid for all farms and all crops (including set-aside, but excluding some permanent crops). Also, decoupled CAP animal support, based on 2000–02 production, will be paid for individual farms; however, 69% of bull premia and 100% of suckler cow premia will remain coupled to production, i.e. paid per animal. The sum of coupled bull and suckler cow premia will not exceed 75% of the bull premia paid in the reference period 2000–02. Overall, 85% of CAP support will be decoupled. The farm-specific payments of decoupled animal support will later be included in the flat-rate payment. The timetable of the shift of farm-specific top-ups into the flat rate is still open (MAF, 2005).

Receiving decoupled CAP support will not require any agricultural commodity production. Yet farmland has to be kept in good agricultural and environmental condition and this means in practice that land has to be either cultivated or kept as set-aside land. In the REF scenario, no change in the EU level of cereal prices is assumed. The reform of the milk sector will be more radical than that agreed in Berlin in 1999. The intervention price for butter is reduced by 25% (-7% in 2004, 2005 and 2006, and -4% in 2007), which is 10% more than that agreed in Agenda 2000. For skimmed milk powder (SMP), prices will be cut by 15% as agreed in Agenda 2000 (but this will occur in steps of 5% over three years from 2004–06). In 2007, it is assumed that the overall decrease in the average producer price of milk at the EU level will be 16% down from the 2003 price level, i.e. 4 percentage points more relative to the baseline. The price cuts will be compensated by a direct payment of €35.50 per tonne of milk quota. This payment becomes fully decoupled in 2007. Furthermore, 5% of all direct EU payments will be cut (modulated) from 2007.

As regards multifunctionality issues and indicators, the question here is to evaluate to what extent the 2003 CAP reform, notably the decoupling of agricultural income-support direct aids (with the possibility for each member state to maintain some of these direct aids as coupled support) is likely to have positive or negative effects on multifunctionality indicators.

7.3.3   Environmental cross-compliance (ECC)

The environmental cross-compliance (ECC) scenario is identical to the REF scenario, except that one assumes that each member state chooses the full decoupling option and that field edges have to be expanded by 300%, i.e.

from 0.5 to 2 metres wide. The idea of this scenario is to test the contribution of field edges to the habitat index and the possible effects on production.

7.3.4 Attack on domestic support scenario (RED)

The attack on domestic support (RED) scenario assumes, in addition to the ECC scenario that the EU (and Finland) is forced to agree a 20% cut in existing decoupled payments by 2013. By comparing the RED and ECC scenarios, one should be able to say something about the 'relative' efficiency of decoupling as regards multifunctionality.

7.4 Indicator results from the model

7.4.1 Continuation of Agenda 2000

The baseline run of the agricultural sector model indicated (with certain exceptions) that if the Agenda 2000 policy continued, there would be no substantial changes in production volumes. For example, dairy production would remain almost unchanged in all regions. But the total amount of cultivated area, including fallow and cultivated grassland, would fall significantly. The most important change therefore concerns the amount of marginal farming land taken out of production, an area that would add 10-15% to all agricultural land. Such a change results mainly from investments in larger dairy facilities, which in turn lead to a regional concentration of agricultural commodity production within each individual region studied. Thus, the demand for feed (grain and grass) decreases in many areas. This decline weakens endogenous market prices and the profitability of grain production. Because pork and poultry production also continue to become concentrated into large production units, some agricultural land is left idle in relatively less favourable agricultural areas. The relative increase in the uncultivated land area will be largest in northern Finland, but the absolute changes are largest in southern and central Finland. Soil quality is highly heterogeneous even in southern Finland. This heterogeneity is partly taken into account since there are 17 production regions in the DREMFIA model. Nevertheless, soil quality is considered homogenous in each region.

7.4.2 Agenda 2000 vs. the ongoing reform of the CAP

In the baseline scenario, milk production remains relatively stable in all regions since milk payments compensating the price reductions of butter and skimmed milk powder are tied to milk quotas. In the CAP reform scenario, the linkage between milk payments and milk quotas is removed. According to the model results, CAP reform is not likely to result in any drastic decline of agricultural production in Finland, on the aggregate. Milk production may reduce substantially, however, in northern Finland, where dairy products dominate. Some decline of production may also take place in central Finland but the decline of production in these areas will be partly compensated by an increase in production in the Ostrobothnia region, which may benefit from the decreasing values of milk quotas under CAP reform. Hence, some milk quotas from central Finland shift to Ostrobothnia. Since milk quotas cannot be sold from southern Finland to Ostrobothnia (the movement of milk quotas has been restricted in three major quota-trading areas, where the number of sold quotas must equal the number of purchased quotas), the decreased value of milk quotas in southern Finland will facilitate a recovery of milk production from 2010. In northern Finland, which constitutes a third of the quota-trade area, however, not even the reduced milk quota values are sufficient to lead to later recovery in the milk production volume (Figure 7.3).

Figure 7.3 Total milk production volumes in Finland (million litres)

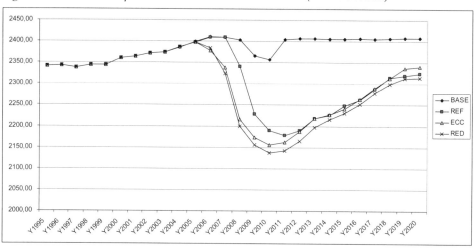

Notes: BASE = Agenda 2000; REF = Luxembourg 2003 reform with national adaptations; ECC = environmental cross-compliance; RED = reduction of CAP payments by 20% by 2013.

This difference results from cuts in the milk price and decoupled CAP payments, which considerably reduce incentives to invest in milk production in the REF scenario. Since many farms are small and production costs are high, most dairy farmers who exit milk production make only the minimum effort to receive the CAP payments, i.e. they leave their land as set-aside. The reduction in overall production volume, on the other hand, provides opportunities for expanding dairy farms (Table 7.2).

Table 7.2 Changes in milk production by 2015 in different scenarios compared with the baseline (in %)

	REF			ECC			RED		
	2010	2015	2020	2010	2015	2020	2010	2015	2020
Southern Finland	-11.1	-9.3	-0.9	-14.8	-9.2	+1.0	-14.4	-8.5	+0.5
Ostrobothnia	-4.0	+3.4	+3.7	-2.2	+3.6	+4.9	-3.3	+3.3	+6.1
Central Finland	-5.9	-11.3	-8.9	-9.1	-13.4	-11.8	-9.3	-14.7	-11.7
Northern Finland	-7.4	-18.6	-22.2	-6.5	-16.5	-17.5	-9.1	-20.5	-25.0
Whole country	-7.1	-6.6	-3.4	-8.5	-6.9	-2.8	-9.3	-7.3	-3.9

Notes: BASE = Agenda 2000; REF = Luxembourg 2003 reform with national adaptations; ECC = environmental cross-compliance; RED = reduction of CAP payments by 20% by 2013

Source: Authors' calculations.

A decreasing number of dairy cows and the partial or full de-coupling of CAP headage payments will gradually result in decreasing beef production and grass area in all parts of Finland, but the effects of CAP reform on pork and poultry production will be minor.

When comparing the agricultural land-use predictions of the ongoing CAP reform scenario to the corresponding results of the extended Agenda 2000 scenario, we found that the REF scenario resulted in a larger green fallow area by almost four-times that of the BASE scenario. Correspondingly, the areas devoted to barley, oats and grass would be significantly smaller under the REF scenario (Table 7.3). Only the most feasible areas of earlier grasslands would be used for grain production. In relative terms, the difference in the green set-aside area between the two scenarios was largest in northern and central Finland. In both regions, the green set-aside area would increase significantly as a result of the REF scenario.

Table 7.3 *Changes in land use by 2015 in different scenarios compared with the baseline scenario (per 1,000 ha) (in %)*

	Cereals area			Grass area			Green set-aside		
	REF	**ECC**	**RED**	**REF**	**ECC**	**RED**	**REF**	**ECC**	**RED**
Southern Finland	-12.8	-13.1	-11.9	-20.4	-23.2	-20.1	+350	+231	+341
Ostrobothnia	-23.2	-25.6	-23.9	-5.2	-2.7	-0.0	+156	+78.7	+143
Central Finland	-45.5	-48.3	-46.8	-11.2	-12.8	-12.6	+440	+357	+450
Northern Finland	+150	+191	+166	-9.7	-13.3	-13.1	+1073	+705	+1,089
Whole country	-19.2	-20.2	-18.9	-11.8	-12.3	-11.1	+305	+206	+299

Notes: BASE = Agenda 2000; REF = Luxembourg 2003 reform with national adaptations; ECC = environmental cross-compliance; RED = reduction of CAP payments by 20% by 2013.

Source: Authors' calculations.

Compared with the baseline scenario, the decoupling of CAP support from production[2] slightly decreases the area under cereals production in southern Finland. The changes in the dairy sector are clearly seen in the proportion of grassland area, which would be approximately 20% smaller in 2015 as a result of the REF scenario. Instead, the fallow area may be over three times larger than that under Agenda 2000. These changes in grassland and fallow areas are also significant in absolute terms, since over a half of the total agricultural area is located in southern Finland.

Ostrobothnia is the second largest agricultural area. If CAP supports were decoupled from production, the area under cereals in 2015 would be reduced by over 20% compared with the baseline scenario. Fallow area, in turn, would be almost 2.5 times larger. If CAP support were disconnected

---

[2] In the REF scenario, 85% of CAP supports are decoupled, and in the ECC and RED scenarios, 100% of CAP supports are decoupled.

from production, the cereals area would decline (relatively) the most in central Finland. In 2015 it would shrink by nearly 50% as a result of the Agenda 2000 policy. The grassland area in turn would be reduced by over 10%, but the fallow area may be almost 10 times larger.

The share of agricultural land under grain was about 9% and the share of set-aside was around 4% in northern Finland in 2003. In the BASE scenario, the share of grain reduces further to 3% by 2015. As opposed to the other regions, the ongoing CAP reform would not reduce the cereals area of northern Finland. Instead the grain area would remain close to the 2003 level, i.e. grasslands (but not grain areas) would be converted to set-aside areas. When many northern dairy farmers exit unprofitable dairy production, this not only adds set-aside areas, but may also lead to an increase in grain areas on those former grasslands where the costs of feed-grain cultivation can be covered. The greatest increase, however, would be in the fallow area, which would be over 10 times larger as a result of the REF scenario. The area under grass, which already covers close to 90% of agricultural land in northern Finland, would be cut by approximately 10% compared with the baseline scenario. While dairy production would fall by 15-20%, this means that production would become relatively more extensive.

The decreasing cereals area also means a decreasing area under pesticide application. With the exception of northern Finland, the chemical pesticide application area across Finland would be smaller under CAP reform scenarios than as a result of the baseline scenario, since cereal areas would decrease if direct aid payments were decoupled from production. If we examine the land-use results at the whole country level, the pesticide application areas would be the largest under the baseline scenario and the smallest under the RED scenario. This would benefit farmland birds for example, since the reduced use of pesticides may increase the amount of insect prey.

The REF scenario would result in lower nitrogen and phosphorus surpluses only temporarily in all regions. Regional concentrations of dairy and beef production would be stronger in the REF scenario compared with the BASE scenario (Table 7.4). This effect would in turn drive up the nutrient balances again. Significant regional concentration of dairy production and larger farms imply more intensive grassland management, despite lower milk prices due to CAP reform. For this reason, the nutrient balance would increase even in central Finland, where dairy production volume would decrease slightly. The high increases of the average nutrient balances in central Finland would partly stem from a drastically diminishing grain area and more intensive dairy production. In central Finland there would be some scarcity of land available for large dairy farms, which would drive up the nutrient balances. The nitrogen and phosphorous balances would remain below the baseline scenario levels in southern Finland and also in northern Finland, where the overall milk production volume would decrease considerably, and there would be less pressure for intensive dairy production (Table 7.4).

Table 7.4 Changes in the aggregate nitrogen and phosphorus balance (kg/ha) by 2015 in different scenarios compared with the baseline (in %)

	Aggregate nitrogen balance			Aggregate phosphorus balance		
	REF	ECC	RED	REF	ECC	RED
Southern Finland	-17.8	-18.2	-18.5	-2.2	-3.5	-2.3
Ostrobothnia	-7.2	-2.5	-0.0	+5.8	+11.2	+8.1
Central Finland	+10.1	+10.2	+11.1	+19.7	+20.1	+18.7
Northern Finland	-0.1	-6.0	-1.8	-6.9	-7.5	-6.5

Notes: BASE = Agenda 2000; REF = Luxembourg 2003 reform with national adaptations; ECC = environmental cross-compliance; RED = reduction of CAP payments by 20% by 2013

Source: Authors' calculations.

The above-mentioned changes in land allocation would lead to a slightly more uneven, aggregate land-cover class distribution in southern and central Finland and in Ostrobothnia (see Shannon's diversity index in Table 7.5). In northern Finland, the value of SHDI would slightly increase in CAP reform scenarios along with the higher cereals and uncultivated agricultural areas. Increased uncultivated area results in increased SHDI values in the CAP support-reduction (RED) scenario. Expanded field edges in the ECC scenario provide a further increase of diversity in land use. On the other hand, a reduction of support from the common agricultural policy does not lead to a decrease in the SHDI in comparison with the REF scenario.

Table 7.5 Changes in Shannon's diversity index and habitat index by 2015 in different scenarios compared with the baseline (in %)

	SHDI			Habitat index		
	REF	**ECC**	**RED**	**REF**	**ECC**	**RED**
Southern Finland	-13.7	-6.0	-8.8	+53.7	+30.8	+52.2
Ostrobothnia	-5.6	-2.1	-4.4	+30.0	+12.4	+29.6
Central Finland	-17.7	-7.5	-17.7	+69.5	+52.2	+70.3
Northern Finland	+20.0	+36.4	+24.4	+46.6	+27.2	+43.3

Notes: BASE = Agenda 2000; REF = Luxembourg 2003 reform with national adaptations; ECC = environmental cross-compliance; RED = reduction of CAP payments by 20% by 2013

Source: Authors' calculations.

In order to evaluate changes in biodiversity, a habitat index has been calculated. While the baseline scenario shows gradually decreasing levels of the habitat index on agricultural lands (as the grass area declines), the rapid expansion of green set-aside in all other scenarios results in a significant increase in the habitat index. This effect occurs because green set-aside is considered a more valuable living environment than grain crops by almost five times for certain indicator species (in this case, butterflies). Nevertheless, if large areas are idled and gradually converted to forest, the biodiversity value of agricultural lands would diminish. The actual effect on biodiversity in the overall ecosystem (comprising agricultural land, forests, ponds, lakes, etc.) where farmlands are idled, however, is uncertain. Hence the habitat index calculated only shows the value of different uses of agricultural land.

According to the model results, agricultural income, as well as agricultural income per labour hour, is higher in the CAP reform scenarios. In the case of partial or full de-coupling, a farmer may reduce relatively less-profitable activities without losing all support. This effect is a commonly perceived one of CAP reform and a motivation for it. Yet according to the model results, aggregate agricultural income would decrease in northern Finland owing to diminishing dairy production. This result is understandable since in the north there are few alternatives to dairy and beef production (Figure 7.4 and Table 7.6).

Figure 7.4 Agricultural incomes in Finland (€1000)

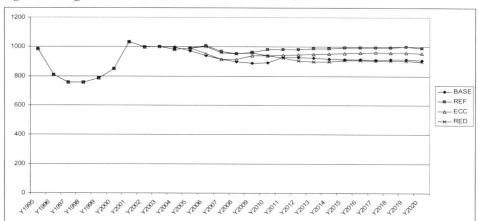

Notes: BASE = Agenda 2000; REF = Luxembourg 2003 reform with national adaptations; ECC = environmental cross-compliance; RED = reduction of CAP payments by 20% by 2013.

Source: Authors' calculations.

Table 7.6 Changes in total agricultural income and income per hour of labour by 2015 in different scenarios compared with the baseline (in %)

	Agricultural income (€ million)			Agricultural income per hour (€/hour)		
	REF	ECC	RED	REF	ECC	RED
Southern Finland	+12.6	+7.9	+2.1	+16.8	+14.5	+5.4
Ostrobothnia	+6.9	+3.0	-2.1	+10.4	+7.0	+0.6
Central Finland	+5.3	+2.7	-4.6	+17.4	+18.2	+9.5
Northern Finland	-3.1	-6.4	-6.6	+13.4	+11.8	+10.7
Whole country	+8.5	+4.5	-0.1	+14.9	+12.3	+4.9

Notes: BASE = Agenda 2000; REF = Luxembourg 2003 reform with national adaptations; ECC = environmental cross-compliance; RED = reduction of CAP payments by 20% by 2013.

Source: Authors' calculations.

In any scenario, including the BASE scenario, agricultural employment is set to fall in all the regions of Finland. This decline stems from the fact that traditionally, farm size has been relatively small in Finland for historical reasons. In spite of the relatively rapid decline of cattle farms over the last 10 years there is still substantial scope for growth in farm-size. Looking towards the future, the decline in agricultural employment is likely to be greatest in Finland's northern region and least in the southern region and in Ostrobothnia (see Figure 7.5 and Table 7.7).

Figure 7.5 Direct and indirect labour in agriculture in Finland in the BASE scenario (per 1,000 employees)

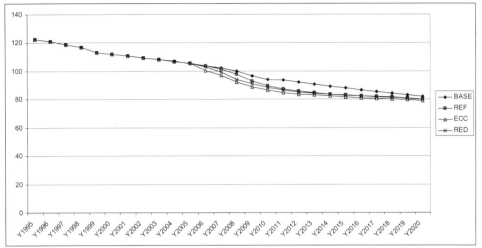

Source: Authors' calculations.

Table 7.7 Changes in the direct and indirect labour of agriculture (per 1,000 employees) in different scenarios compared with the baseline (in %)

	REF			ECC			RED		
	2010	2015	2020	2010	2015	2020	2010	2015	2020
Southern Finland	-3.5	-3.6	+3.8	-9.6	-5.7	+1.7	-5.1	-3.1	+4.5
Ostrobothnia	-7.2	-3.1	-3.0	-6.4	-3.7	-2.8	-6.5	-2.7	-1.1
Central Finland	-3.5	-10.3	-8.4	-8.0	-13.1	-11.5	-7.2	-12.9	-12.2
Northern Finland	-3.1	-14.6	-20.1	-5.5	-14.7	-17.2	-5.7	-15.6	-21.1
Whole country	-4.6	-5.6	-2.4	-8.1	-7.3	-3.7	-6.0	-5.9	-2.4

Notes: BASE = Agenda 2000; REF = Luxembourg 2003 reform with national adaptations; ECC = environmental cross-compliance; RED = reduction of CAP payments by 20% by 2013.

Source: Authors' calculations.

The food industry in Finland has also experienced major structural changes in the last 10 years. These changes are still going on. Hence it is assumed in this study that indirect employment in agriculture is considered to remain fixed for the agricultural labour force. For example, the transportation of inputs and outputs will employ less labour as the number of farms decrease. Overall, the efficiency of labour will increase significantly and in the same magnitude in agriculture and in upstream and downstream industries. This means that the average reduction in agricultural employment (-18% in the baseline and -23% in the CAP reform scenarios by 2015) would be accompanied by the same change in indirect employment. This assumption will be relaxed as soon as new input–output data in each of the 20 provinces is obtained.

Summary and conclusions

The aim of this study is to predict and compare the multifunctionality effects of various agricultural policy reforms in Finland. Since the multifunctional value of agriculture lies in the joint production process of agricultural and public goods alongside other externalities, the impact of different agricultural policies on production, biological and employment factors have been analysed. The study has applied agricultural sector modelling as the research method, which takes into account changes in the profitability of different agricultural production lines and the resulting changes in land use and production intensity. Since the model does not explicitly consider the links between agriculture and the national economy, the effects on direct and indirect employment in agriculture are considered external to the model.

CAP reform, possibly through the partial or full-decoupling of CAP payments, is not likely to result in any drastic decline of agricultural production in Finland on the whole. Yet according to the economic analysis here, milk production may reduce substantially in northern Finland where dairy farming is the dominant line of production. Some decline of production may also take place in central Finland, but decreasing production in these areas will be partly compensated by increasing production in the Ostrobothnia region, which may benefit from the falling values of milk quotas in CAP reform.

A decreasing number of dairy cows and the partial or full de-coupling of CAP headage payments would gradually result in reduced beef production and grasslands in all regions of Finland. Furthermore, the enlarging size and regional concentration of dairy production is likely to keep up the nutrient balances on agricultural land despite falling milk prices, which, *ceteris paribus*, would reduce the intensity of milk production.

The scarcity of land in the relatively most-competitive areas may even increase the phosphorous balance – which is a risk in terms of nutrient runoffs and water quality.

In any scenario (including the baseline) agricultural employment is set to decline significantly throughout Finland. It is notable that the agricultural labour force is likely to shrink substantially irrespective of agricultural policy. According to our results, only a significant reduction in agricultural supports would speed up the decline in agricultural labour in southern Finland and Ostrobothnia, given that overall animal production will in any scenario stay at the present level or gradually rise in these relatively more competitive regions. But in northern Finland, CAP reform would reduce agricultural employment (and related indirect employment) significantly because of the substantial reduction of milk production, which is the dominant line of production in this part of the country. Interestingly, in the ECC scenario the reduction of agricultural employment would be less than that under the REF scenario. In addition to restrictions in milk-quota trading among the regions, this result is partly owing to enlarged field edges, which slow down the concentration of production in other areas along with the outflow of milk quotas from northern Finland to some parts of central Finland. Hence the enlarged field edges would slightly mitigate the decline in dairy production and employment in northern Finland. From a multifunctionality viewpoint, this is an interesting result as environmental cross-compliance may also enhance socio-economic viability as measured by agricultural labour.

On the effects of policies on agricultural land use, the main finding is that the amount of fallow land (especially green fallow) would increase considerably if agricultural support payments are decoupled from production. Although the expenses of establishing green fallows are higher than for bare fallows, the maintenance costs of green fallows are less. Based on the farm-level production cost calculations of the Union of Rural Advisory Centres (MKL, 1995), green fallows are more profitable than bare fallows over a five-year period, and thus the predicted rise in the area of green fallows is justified. Nevertheless, there is substantial uncertainty about the number of green fallow areas in the future, since at the farm level the choice of set-aside management also depends on the opportunity cost of labour and the age of production capital.

In addition, it should be noted that the above results depend on the environmental cross-compliance requirement of keeping the land in good agricultural condition. Without this requirement, the decoupling of support payments could in turn lead to land abandonment.

A significant reduction of agricultural support is another factor that may also prompt land abandonment. This effect can be seen in the 'attack on domestic support' scenario where CAP payments are reduced by 20%. As a result, idled land grows to nearly 25% of the total agricultural land area. Since agricultural land comprises only 8% of all land in Finland, one could argue that any reduction in agricultural land would mean a loss in biodiversity or at least diversity at the landscape level.

The habitat index calculated does not assign any value to idled or afforested agricultural land but solely considers agricultural land. Yet the calculated habitat index does show that in any policy scenario the remaining cultivated agricultural land would become biologically richer owing to an increase in green set-aside areas. Expanded field edges, required in the cross-compliance scenario, would provide significantly richer habitats for various species, such as butterflies. Since an increased set-aside area is a major expected outcome of CAP reform, providing sufficient incentives for green set-aside areas would make it possible to attain a higher level of biodiversity. On the other hand, our results show that changes in field edges (a change from 0.5 to 2 metre-wide edges were studied) do not imply any rise in the overall habitat value of agricultural lands. On the contrary, the enlarged field edges result in a higher intensity of grassland cultivation and larger grain areas in some places in which dairy and beef are dominant lines of production. In our calculations, the higher intensity effectively lowers the habitat value of grasslands.

Consequently, wider field edges may even reduce the overall habitat index, especially if field parcels are small. This result, however, as well as the calculation procedure of the habitat index, needs to be discussed with environmental scientists.

At the landscape level, these policy reforms, in which support is decoupled, change land use and decrease the diversity of agricultural land-cover classes in almost all parts of the country, except in northern Finland. But the effect on biological diversity may not be equal to changes in Shannon's diversity index, since at the species level, green fallows seem to have some positive effects, especially on the densities and abundance of farmland birds (Haukioja et al. 1985; Helenius et al. 1995; Tiainen & Pakkala 2000; Tiainen & Pakkala, 2001). Firbank et al. (2003) concluded that rotational set-aside in particular provides suitable habitats for breeding birds, but the benefits of short-term set-aside for arable plants in England were small. Corbet (1995), on the other hand, considered long-term set-aside a possibility for establishing patches of undisturbed perennial herbaceous vegetation and their associated fauna. Steffan-Dewenter & Tscharntke (1997), Critchley & Fowbert (2000) and Kuussaari & Heliölä (2004) remarked that green fallows are poorer habitats than meadows when considering the species diversity of vascular plants or butterflies and other insects.

Further research work and new data are necessary to evaluate the employment effects. Closer cooperation with environmental scientists is needed in the development of the habitat index, especially as production techniques (such as no-till cultivation) are changing. In addition, different habitat indexes may be constructed in order to separately quantify the richness of animal and plant species. Interdisciplinary work is also important in terms of the overall credibility of the economic modelling approach in evaluating the multifunctionality issue: the material flows and production biological relationships in the sector model need to be regularly updated with new research and data at the farm level. Hence the assessment of different agricultural policies on many aspects of multifunctionality is a continuous research agenda rather than a single project providing fixed coefficients for future indicator calculations. The credibility of the production economics and biological relationships of the economic model determine the validity of the indicator results. The overall framework and economic logic of microeconomic simulation models provides a consistent assessment of the many aspects of multifunctionality in comparison with a number of different expert opinions.

Bibliography

Anderson, K. (2000), "Agriculture's Multifunctionality and the WTO", *Australian Journal of Agricultural and Resource Economics*, Vol. 44, No. 3, pp. 475-94.

Armington, P. (1969), *A Theory of Demand for Products Distinguished by Place of Production*, IMF Staff Paper No. 16, IMF, Washington, D.C., March, pp. 159-78.

Boisvert, R. (2001), "A note on the concept of jointness in production" in OECD *Multifunctionality: Towards an analytical framework*, Technical Annexes (Annex 1, pp. 105-23, Annex 2, pp. 125-32), OECD, Paris.

Bäckman, S., S. Vermeulen and V.-M. Taavitsainen (1997), "Long-term fertilizer field trials: Comparison of three mathematical response models", *Agriculture and Food Science in Finland*, Vol. 6, pp. 151-60.

Corbet, S.A. (1995), "Insects, plants and succession: Advantages of long-term set-aside", *Agriculture, Ecosystems and Environment*, 53, pp. 201-17.

Cox, T.L. and J.-P. Chavas (2001), "An interregional analysis of price discrimination and domestic policy reform in the US dairy sector", *American Journal of Agricultural Economics*, Vol. 83, No. 1, pp. 89-106.

Critchley, C.N.R. and J.A. Fowbert (2000), "Development of vegetation on set-aside land for up to nine years from a national perspective", *Agriculture, Ecosystems and Environment*, 79, pp. 159-74.

Duelli, P. (1997), "Biodiversity evaluation in agricultural landscapes: An approach at two different scales", *Agriculture, Ecosystems and Environment*, 62, pp. 81-91.

European Commission (1999), *Reform of the Common Agricultural Policy (CAP)*, Brussels (retrieved from http://europa.eu.int/scadplus/leg/en/lvb/l60002.htm).

————— (2003), Explanatory memorandum: A long-term perspective for sustainable agriculture, 5586/03 COM(2003) 23 final, Brussels.

Farm Register (2002), Agriculture, Forestry and Fisheries 2003:60, Official Statistics of Finland, Information Centre of the Ministry of Agriculture and Forestry, Helsinki (retrieved from http://tike.mmm.fi).

Firbank, L.G., S.M. Smart, J. Crabb, C.N.R. Critchley, J.W. Fowbert, R.J. Fuller, P. Gladders, D.B. Green, I. Henderson and M.O. Hill (2003), "Agronomic and ecological costs and benefits of set-aside in England", *Agriculture, Ecosystem and Environment*, 95, pp. 73-85.

Guyomard, H. and F. Levert (2001), *Multifunctionality, trade distortion effects and agricultural income support: A conceptual framework with free entry and land price endogeneity*, paper presented at the seminar "Multifunctional Agriculture", held in Bergen, Norway, in February.

Guyomard, H., C. Le Mouel and A. Gohin (2004), "Impacts of alternative agricultural income support schemes on multiple policy goals", *European Review of Agricultural Economics*, 31, pp. 125-48.

Haukioja, M., P. Kalinainen and K. Nuotio (1985), *Maatalouden vaikutus peltolinnustoon: esitutkimusraportti*, Ympäristöministeriön ympäristön- ja luonnonsuojeluosaston julkaisu A:34, Ympäristöministeriö, Helsinki.

Heckelei, T., P. Witzke and W. Henrichsmeyer (2001), "Agricultural Sector Modelling and Policy Information Systems" in *Proceedings of the 65th European Seminar of the European Association of the Agricultural Economics (EAAE)*, 29-31 March 2000, Bonn, Germany.

Heikkilä, A.-M., L. Riepponen and A. Heshmati (2004), *Investments in new technology to improve productivity of dairy farms*, paper presented at the 91st EAAE Seminar "Methodological and Empirical Issues of Productivity and Efficiency Measurement in the Agri-Food System", held in Rethymnon, Greece on 24-26 September.

Helenius, J., S. Tuomola and P. Nummi (1995), "Viljely-ympäristön muutosten vaikutus peltopyyn ravintoon", *Suomen Riista*, 41, pp. 42-52.

Hietala-Koivu, R., J. Lankoski and S. Tarmi (2004), "Loss of biodiversity and its social cost in an agricultural landscape", *Agriculture, Ecosystem and Environment*, 103, pp. 75-83.

Hietala-Koivu, R., L. Tahvanainen, I. Nousiainen, T. Heikkilä, A. Alanen, M. Ihalainen, L. Tyrväinen and J. Helenius (1999), *Visuaalinen maisema maatalouden ympäristöohjelman vaikuttavuuden seurannassa*, Maatalouden tutkimuskeskuksen julkaisuja, Sarja A 50, Maatalouden tutkimuskeskus, Jokioinen, Finland.

Jeanneret, Ph., B. Schüpbach and H. Luka (2003), "Quantifying the impact of landscape and habitat features on biodiversity in cultivated landscapes", *Agriculture, Ecosystem and Environment*, 98, pp. 311-20.

Jensen, H.G. and S.E. Frandsen (2003), *Impacts of the Eastern European Accession and the 2003 Reform of the CAP*, FOI Working Paper No. 11/2002, Danish Research Institute of Food Economics, Copenhagen.

Knuuttila, M. (2004), "Elintarvikesektorin työllisyysvaikutukset – Panos-tuotosanalyysi maakunnittain", *Maa- ja elintarviketalous* 56: 87 s (retrieved from http://www.mtt.fi/met/pdf/met56.pdf).

Kuussaari, M. and J. Heliölä (2004), "Perhosten monimuotoisuus eteläsuomalaisilla maatalousalueilla", in M. Kuussaari, J. Tiainen, J. Helenius, R. Hietala-Koivu and J. Heliölä (eds), *Maatalouden ympäristötuen merkitys luonnon monimuotoisuudelle ja maisemalle: MYTVAS-seurantatutkimus 2000-2003,* Suomen ympäristö (in press).

————— (2004), "Butterfly diversity in agricultural landscapes of southern Finland" (in Finnish), *Suomen ympäristö*, 709.

Lankoski, J. and M. Ollikainen (2003), "Agri-environmental externalities: A framework for designing targeted policies", *European Review of Agricultural Economics*, 30, pp. 51-75.

Lankoski, J., E. Lichtenberg and M. Ollikainen (2004), *Performance of alternative policies in addressing environmental dimensions of multifunctionality*, Discussion Paper No. 4, Department of Economics and Management, University of Helsinki.

Lehtonen, H. (2001), "Principles, structure and application of dynamic regional sector model of Finnish agriculture", PhD thesis, Publication 98, Systems Analysis Laboratory, Agrifood Research Finland, Economic Research (MTTL), Helsinki University of Technology.

————— (2004), "Impacts of de-coupling agricultural support on dairy investments and milk production volume in Finland", *Acta Agriculturae Scandinavica, Section C: Food Economics,* Vol. 1. No. 1, April, pp. 46-62.

MAF (2005), "Tilatukijärjestelmän sisällöstä neuvottelutulos", Press Release 1 March 2005, Ministry of Agriculture and Forestry (retrieved from http://www.mmm.fi/tiedotteet/index.asp?ryhma=1).

McGarigal, K. and B.J. Marks (1995), *FRAGSTATS: Spatial Pattern Analysis Program for Quantifying Landscape Structure,* PNW-GTR-351, USDA Forest Services, Portland, OR.

MKL (1995), *Mallilaskelmat,* Suunnitteluosaston sarja A. Maaseutukeskusten liitto, Helsinki.

OECD (2001), *Multifunctionality: Towards an analytical framework,* OECD, Paris.

Ollikainen, M. and J. Lankoski (2004), *Multifunctional agriculture: The effect of non-public goods on socially optimal policies*, paper presented at the 90th EAAE Seminar, "Multifunctional Agriculture, Policies and Markets: Understanding the Critical Linkage", held in Rennes on 28-29 October.

Paarlberg, P., M. Bredahl and J. Lee (2002), "Multifunctionality and agricultural trade negotiations", *Review of Agricultural Economics*, 24, pp. 322-35.

Peterson, J., R. Boisvert and H. de Gorter (2002), "Environmental policies for a multifunctional agricultural sector in open economies", *European Review of Agricultural Economics*, 29, pp. 423-43.

Pro Agria (2005), *Production results from dairy farms*, Central Union of Rural Advisory Centres in Finland (retrieved from http://www.proagria.fi/palvelut/tuotantotulokset/maito.asp).

Romstad, E., A. Vatn, P.K. Rørstad and V. Søyland (2000), *Multifunctional Agriculture: Implications for Policy Design*, Agricultural University of Norway, Ås.

Sairanen, A., J. Nousiainen and H. Khalili (1999), "Korkean väkirehumäärän vaikutus maitotuotokseen ja tuotannon kannattavuuteen", Poster paper (P7) presented at Agro-Food '99 2-4 February 1999, Tampere (in Finnish).

———— (2003), "Milk yield responses to concentrate supplementation on pasture" in O. Niemeläinen and M. Topi-Hulmi (eds), *Proceedings of the NJF's 22nd Congress of Nordic Agriculture in a Global Perspective*, 1-4 July 2003, Turku, Jokioinen: MTT Agrifood Research Finland, NJF, p. 23 (retrieved from http://portal.mtt.fi/pls/portal30/docs/folder/agronet/yhteiset hankkeet/njf/njf2003/2.pdf).

Soete, L. and R. Turner (1984), "Technology diffusion and the rate of technical change", *The Economic Journal*, 94, pp. 612-23.

Steffan-Dewender, I. and T. Tscharntke (1997), "Early succession of butterfly and plant communities on set-aside fields", *Oecologia*, 109, pp. 294-304.

Vatn, A. (2002), "Multifunctional agriculture: Some consequences for international trade regimes", *European Review of Agricultural Economics*, Vol. 29, pp. 309-27.

Tiainen, J. and T. Pakkala (2000), "Population changes and monitoring of farmland birds in Finland" in *Linnut-vuosikirja 1999*, Helsinki: BirdLife Suomi, pp. 98-105.

———— (2001), "Birds" in M. Pitkänen and J. Tiainen (eds), *Biodiversity of Agricultural Landscapes in Finland*, BirdLife Finland Conservation Series, No. 3, Helsinki, pp. 33-50.

8. Modelling agricultural multifunctionality
A case study of the Czech Republic

Tomáš Doucha and Ivan Foltýn

Introduction

The transformation of the Czech agricultural sector and its adjustment to new social and economic conditions has been ongoing since 1990, after the velvet revolution. The substance of the transformation consists of the restitution of ownership rights and titles for agricultural assets (property transformation) and of the restructuring of farms, their production and land use. Since the latter half of the 1990s, the restructuring has also been linked to the development of the multifunctionality of Czech agriculture. This orientation has gained importance since the accession of the Czech Republic to the EU in 2004, the entry of Czech farms into the EU Single Market under the conditions of the common agricultural policy (CAP) and the application of the European model of agriculture.

In section 8.1, this chapter summarises the actual state of the Czech agricultural sector after 15 years of transformation, with the stress placed on its multifunctional characteristics and on the conditions influencing multifunctionality. Applying a non-linear optimising model and using data from the Farm Accountancy Data Network (FADN) and other parameters (in section 8.2), the effects of defined policy scenarios are simulated up to 2010 (in section 8.3) on the production structure and multifunctionality of Czech agriculture (in section 8.4). The effects on individual farm categories are projected, differentiated by regional aspects and their expected behavioural formulas and responses to policy stimuli. The most important policy issues arising from the simulations are presented in the conclusions.

The multifunctionality of agriculture – its positive externalities as public goods related to the environment and to rural development – is defined according to the methodology presented in Guyomard (2006) and in Dwyer et al. (2005). The indicators of multifunctionality are also derived from the referenced ENARPRI Working Papers, which can be adjusted to the possibilities of the applied mathematical model. The same point applies to the definitions of the policy scenarios.

8.1 The multifunctionality of Czech agriculture – Current status

8.1.1 Czech agricultural policy

Multifunctionality in the Czech agricultural sector has been developing under the conditions determined by agricultural and other policies and the general institutional framework of society. Up to the mid-1990s, Czech agricultural policy was prevailingly oriented towards property transformation in the sector and income support for the emerging new farm structures. After 1994, the following stages of Czech agricultural policy can be observed, as outlined below (see also Table 8.1).

1995–97: Restructuring

Agricultural policy in the period 1995–97 was characterised by:

- continuing support for restructuring and stabilising the new emerging farm structure (41% of all budgetary support);
- introduction of new support for grassland in 'less-favoured areas' (LFAs) in order to maintain the landscape;
- support for the environment and multifunctionality was mainly through LFA payments, but new support was also given for non-food outputs of agricultural production (mainly for biofuel);
- a higher level of protection for domestic consumers through administrative barriers for exports (cereals, oilseeds, etc.); and
- protection for domestic producers at the general level agreed in the Uruguay round of the GATT (approximately 2-2.5 times lower than EU protection), which was only slightly eroded by bilateral and multilateral trade agreements (e.g. the Central European Trade Agreement).

1998–2003: CAP-like policy

Agricultural policy in the period 1998–2003 was oriented towards a gradual adjustment to the CAP and future EU accession, and included these key aspects:

- a growing total level of support (by more than 60% in nominal terms compared with the previous period), particularly in the category of income support (38% of all budgetary support), based on CAP-like marketing frameworks and measures;

- a growing share of support for the environment and multifunctionality (31% of all budgetary support), with the implementation of (LFA) payments and the continuing high level of support for non-food outputs of agricultural production (biofuel); and

- a decrease in the actual tariff protection through the implementation of new trade agreements with the EU ('double-profit' and 'double-zero' agreements), but with protection levels remaining on a higher level than in the EU.

2004–05: The CAP

Czech agricultural policy in the first years after EU accession (2004–05) has featured the following elements:

- a sharp increase in the total level of budgetary support (by 68% compared with the previous period);

- a prevailing share of income support in the total package of budgetary support (more than 55%). Income support in the form of direct payments has consisted of decoupled SAPS[1] payments and coupled national, complementary, direct payments (the so-called 'top-up payments'). With a high share of coupled top-up payments, all the direct payments have thus functioned as coupled support during this start-up period;

- direct payments conditioned on 'good farming practices', but with reduced enforcement;

---

[1] SAPS refers to the Simplified Administrative Payment Scheme.

- an increase of support for the environment and multifunctionality, with a predominant share of LFA payments and a growing share of other support measures (for biodiversity and rural development), but with a sharp decrease of the budgetary support for biofuel (as a consequence of EU regulations in this sector). Owing to relatively weak payment conditions and other factors, however, in reality LFA payments and other forms of environmental support have been functioning as additional direct payments; and

- entry into the EU Single Market with 'zero' protection for the sector, but with a higher average level of protection against the rest of the world compared with the pre-accession period.

Table 8.1 Annual budgetary supports for Czech agriculture from 1995 to 2005, by policy goals

Goals	1995–97		1998–2003		2004–05	
	CZK (mn)	%	CZK (mn)	%	CZK (mn)	%
Restructuring	4,635	41.38	5,457	30.27	4,878	16.13
Incomes	2,208	19.72	6,780	37.61	16,756	55.42
Environment	2,469	22.04	5,518	30.61	7,993	26.44
Consumers	1,888	16.86	271	1.50	609	2.01
Total	*11,199*	*100.00*	*18,025*	*100.00*	*30,235*	*100.00*
Of which environment						
Landscape, LFA	1,742	70.54	3,087	55.94	4,595	57.49
Water, soil	20	0.82	137	2.48	714	8.93
Biodiversity	232	9.40	337	6.10	1,057	13.22
Ecological farming	0	0.00	141	2.55	240	3.01
Forestry, rural[a]	24	0.96	72	1.31	746	9.33
Non-food use	451	18.28	1,745	31.62	642	8.03
Total	*2,469*	*100.00*	*5,518*	*100.00*	*7,993*	*100.00*

[a] This is the only support available under the agricultural policy.

Source: Database of agricultural policy, VUZE.

8.1.2 Farm structure – Land use and ownership

During the entire transformation period, besides the legislation concerning property, the quality of the land market has been one of the most important factors shaping the Czech farm structure and multifunctionality. Owing to path dependencies from the Czech Republic's 'land history' and to ineffective reform instruments (particularly the instruments and financial resources for land consolidation and re-parcelling in cadastres – the elementary official territorial unit in the Czech Republic), the land market has remained undeveloped. The privatisation of state land has proven to be the most significant driving force for its development. Nevertheless, those land users originating from the pre-reform period have retained real power and advantages in the land market. Thus, actual land use and ownership continues to strongly influence the present and future situation in Czech agriculture and its multifunctional roles (see Table 8.2).

Table 8.2 Land users and owners of the Czech utilised agricultural area (2004)

Farms/ owners	State	Muni-cipal	PE[a] farms	LE farms	PP-LE	Other PP	Total (000 ha)	Total (%)	No.	Avg. size (ha)
Subsistence	–	–	40	–	–	–	40	1.11	19,189	0.2
Family	30	5	205	–	–	185	425	11.81	30,231	14.1
Ind.	320	10	65	60	–	1150	1605	44.58	3,704	433.3
CF-M	125	5	–	40	75	395	640	17.78	668	958.1
CF-O	110	5	–	20	180	540	855	23.75	667	1,281.9
Other	35	–	–	–	–	–	35	0.97	180	194.4
Total (000 ha)	620	25	310	120	255	2,270	3,600	100.00	54,639	65.9
Total (%)	17.22	0.69	8.61	3.33	7.08	63.06	100.00	–	–	–
Number	1	6,000	2,000	28,000	50,000	3,000,000	3,086,001	–	–	–
Average size (ha)	620,000	4.17	155.00	4.29	5.10	0.76	1.17	–	–	–

[a] Land leased by PE to other categories of farms is included in OPP (see definitions below).

Notes: PE/LE = physical/legal entity; PP = physical persons; CF-M/O = coops and joint stock companies (M = with a power of management; O = with a power of owners); other companies included in the category of individual farms; PP-LE = PP as members/shareholders of farms

Sources: Czech Statistical Office – Agrocensus 2000, the Czech Land Fund and authors' estimations.

The main conclusions that can be derived from Table 8.2 are as follows:

- There is an extreme concentration of land use (around 5% of the largest farms occupy almost 75% of the utilised agricultural area or

UAA[2]). The dual structure of land use stands against an extreme fragmentation in the land ownership (millions of small owners).

- Czech farms own only about 12–13% of the land, with the remaining agricultural land being leased.

- Family farms only occupy around 13% of the UAA.

- Large individual farms (including partnership farms and limited liability companies) are the most dynamic farm category, occupying nearly half of the UAA at present. Their share in the UAA (also supported by the land privatisation efforts) has been growing. This trend has occurred through the enlargement of family farms and also through formal or informal changes to those collective farms (coops and joint stock companies) in which there has been a concentration of property or economic power in the hands of managers.

- From another point of view, about two-thirds of the UAA is occupied by 'profit-oriented' farms; the remaining one-third is utilised by 'income-oriented' farms, with a stronger self-employment focus.

- Concerning land ownership and use, non-agricultural and foreign capital has been penetrating the sector at an increasing rate in the last few years (as an obvious consequence of the present and expected profitability of the Czech farming/land sector).

Such land use and ownership structures have some implications for the development of multifunctionality in agriculture:

- In principle, there are high transaction costs accompanying any changes in land use or in land ownership. These costs result in passive behaviour on the part of landowners concerning the land market or in serious barriers for land users, e.g. in needed (and therefore supported) conversions of arable land into grassland (landowners block the conversion).

- There is a risk of an extremely high level of diversion of direct payments away from agriculture and from rural areas through land

---

[2] UAA approximately represents the area of Czech agricultural land that is eligible for direct payments. The acreage of the UAA (about 3.5-3.6 million ha) differs from the total acreage of Czech agricultural land (4.3 million ha) based on the registration of ownership plots. Some of the difference (about 300,000 ha) can be considered as abandoned land.

ownership and leased land (today a reasonable number of landowners live in towns). At present, in the Czech Republic the diversion of support is hampered by the low degree of flexibility in the land market, so this is more of a future risk.

- The prevailing profit orientation of farms represents another risk for multifunctionality. The continuing investment support for farm modernisation will clearly lead to a further reduction of labour, without a proper incentive for the establishment of new job opportunities on those farms if new non-agricultural activities are not sufficiently profitable.

8.1.3   Farm categories and their characteristics

The effects of different policy scenarios on multifunctionality are modelled below for individual farm categories. The farm categories are defined through the application of two main criteria:

- the regional location of farms, reflecting also the local share of LFA. Each Czech farm can be identified by its location in the so-called 'production regions', reflecting soil productivity, i.e. in
  - the hilly region (H), which simulates 100% of the LFAs in the area of a farm;
  - the potato region (P), which simulates 50% of the LFAs in the area of a farm;
  - the maize and sugar beet regions (M), which simulate 0% of the LFAs in the area of a farm;
- the behaviour of farms and their expected reflection of policy measures/stimuli, more specifically,
  - profit-oriented farms (P), optimising the rate of profit from inputs/assets;[3] and
  - income-oriented farms (I), optimising the level of gross margin or the maximum profit generated by farming.[4]

---

[3] For modelling, this farm category is represented by large individual farms with more than 300 ha of agricultural land.

[4] For modelling, this farm category is represented by family farms with 50–100 ha of agricultural land.

Combining these two criteria, six categories of farms are recognised for modelling: HP, HI, PP, PI, MP and MI. Based on FADN data and with the conversion of land use and production structures for farms with 100 ha of agricultural land, the main indicators for all selected farm categories are presented in Table 8.3.

Table 8.3 Economic and structural indicators for selected farm categories 2004*

Indicator	Unit	Hilly (H)		Potato (P)		Maize & sugar beet (M)	
		Profit	Income	Profit	Income	Profit	Income
Arable land	ha	12.4	19.6	54.6	71.3	96.7	94.0
– cereals	ha	3.0	14.5	36.9	50.0	66.2	61.9
– oilseeds	ha	1.7	0.0	9.0	7.2	8.9	8.3
– forage	ha	7.6	4.4	4.6	10.9	1.6	5.7
– other	ha	0.1	0.7	4.1	3.2	20.0	18.1
Permanent crops	ha	0.0	0.0	0.0	0.2	0.2	2.9
Grassland	ha	87.6	80.4	45.4	28.5	3.1	3.1
Total UAA	ha	100.0	100.0	100.0	100.0	100.0	100.0
Dairy cows	heads	10.4	8.9	7.0	14.6	1.0	3.8
Suckler cows	heads	7.5	7.2	5.1	5.4	0.3	0.5
Ewes/goats	heads	5.9	0.0	1.6	1.5	0.1	0.9
Pigs	000 CZK	4	8	26	206	74	132
Poultry	000 CZK	0	0	0	3	0	0
Eggs	000 CZK	0	0	0	0	0	1799
Livestock units	LU	32.1	48.7	21.5	43.7	6.1	15.0
Labour	AWU	1.40	2.26	1.40	2.97	1.85	3.02
Production	000 CZK	753	1,247	1,497	2,399	2,362	2,982
– crops	000 CZK	363	484	980	1,202	2,138	2,533
– livestock	000 CZK	375	717	487	1,126	155	381
– other	000 CZK	15	46	30	71	69	68
Interim consumption	000 CZK	691	1164	1077	1644	1549	1915
Depreciation	000 CZK	84	192	125	317	266	349

Table 8.3 cont.

Operational subsidies[a)]	000 CZK	529	657	406	388	329	365
Net value added (NVA)	000 CZK	507	548	826	701	912	1047
Labour costs (hired)	000 CZK	251	40	190	40	294	109
Capital costs	000 CZK	7	35	10	15	22	28
Rents for land	000 CZK	25	38	55	48	122	104
Operational surplus	000 CZK	224	442	446	756	475	838
NVA/AWU	000 CZK	362	242	590	236	493	347
Profitability [b)]	CZK/CZK	1.16	1.07	1.24	1.12	1.15	1.20
Production intensity	000 CZK/ha	7.38	12.01	14.67	23.28	22.93	29.14
Production/ AWU	000 CZK	538	552	1,069	808	1,277	987
Share of non-agri. Prod.	%	2.0	3.7	2.0	3.0	2.9	2.3
Interim consumption/ production	%	91.8	93.3	71.9	68.5	65.6	64.2
Depreciation/ production	%	11.2	15.4	8.4	13.2	11.3	11.7

* Calculated for 100 ha.

[a)] Without production taxes

[b)] (Production+operational subsidies)/(interim consumption + depreciation+labour costs including FWU+capital costs+rents for land).

Source: FADN CZ (2004), VUZE.

8.1.4 *The multifunctionality of Czech agriculture*

The development of multifunctionality in the Czech agricultural sector during the reform, applying selected proxy indicators, is shown in Table 8.4.

Table 8.4 Indicators of multifunctionality – Czech agriculture

Indicator	Unit	1989	1995	2004	Index 2004/1989
Land abandonment	000 ha	300	300	300	100.00
Share of arable land in agricultural land	%	75.00	73.00	71.70	95.60
Share of land threatened by erosion	%	35.00	33.00	33.00	94.29
Share of ecological farming on agricultural land	%	0.00	1.00	5.97	–
Of which on arable land and permanent crops	%	0.00	0.50	7.70	–
Number of cows (dairy and suckler)	000 heads	1248	768	574	45.99
Number of sheep	000 heads	399	80	140	35.09
Number of workers in agriculture	000 pers.	533	222	141	26.45
Share of non-agricultural incomes in total farm incomes	%	30.00	20.00	16.00	53.33

Source: Authors' estimations.

Czech agriculture under the socialist regime was characterised by extremely large collective and state farms and by industrial methods of farming, with heavy negative consequences for the environment and landscape. The share of arable land in the total area reached about 75%, although about two-thirds of the Czech agricultural area now finds itself in regions with worse soil and climatic conditions. This side effect has been observed as typical of the socialist policy of full food self-sufficiency at any cost. The approach towards water in the countryside, however, was the most seriously damaging aspect.

During the last 15 years of the transformation of the Czech agricultural sector, its relations to the environment and landscape have not changed in principle, despite the large financial resources spent on this purpose. Any improvements that have occurred have been enforced by the poor economic conditions affecting farms, leading to a reduction in the consumption of fertilisers, pesticides, etc. The main causes of this situation are

- inappropriate agricultural policies (with opportunity costs fostering the continuation of industrial farming and overweighing the stimuli for change);

- ineffective environmental legislation accompanied by weak enforcement of laws;

- the above-mentioned relations between land users and landowners, generating high transaction costs for needed changes; and

- very slow progress in land consolidation/re-parcelling.

Meanwhile, major changes have occurred in the social functions of farms and in their relations to rural areas. Since 1989, nearly 75% of workers have left farms, being largely absorbed by other sectors. Furthermore, the quality of human capital in the agricultural sector has deteriorated, because mainly younger and more educated workers have exited. Agriculture has stopped representing the main source of rural employment and now the risk of growth in rural unemployment – in view of a further inevitable reduction of labour in primary agricultural production – has been increasing. The risk is all the more serious today, because of a relatively low willingness of farms to create new job opportunities in non-agricultural activities for the released workers. Likewise, other social functions previously provided by farms (nursery schools, canteens, health centres, etc.) have been abolished (with some exceptions). Above all, Czech agriculture, with its prevailing industrial character, still has a tendency to reduce the recreational potential in rural areas, functioning against the needed development of rural tourism.

8.2   Methodology

8.2.1   The FARMA-4 model

Optimal farm behaviour in a system of sustainable development

Definition 1

Farm behaviour under the given natural conditions is *economically optimal* if the farm maximises its profit in the framework of all its possible avenues of production. Indeed, economically-optimal farm behaviour can negatively impact the sustainable development of agriculture (e.g. soil fertility) and the environment.

For modelling the influences of farming on the environment, some indicators that can be quantified and used for measuring the effects on the environment have been selected, as set out below.

- **Ratio of grassland.** This indicator generally characterises exposure to soil erosion and the capacity for water retention.

- **Risk to plants.** Broadly-seeded crops in crop rotation pose a risk of soil erosion.

- **Ratio of organic fertilisers.** This indicator is able to predict the losses of soil diversity in ecosystems and soil erosion, and the washing-off of nitrogen and phosphorus from the surface and underground waters, etc. If there is not a proper circulation of organic mass, then an increased share of industrial fertilisers can be supposed, which causes negative ecological effects.

- **Number of breeding cattle (per head) and other farm animals.** This indicator gives basic information about the production of greenhouse gasses in agriculture.

- **Inputs of energy.** This indicator calculates the consumption of fossil energy by the agricultural sector, compared with the production of renewable energy.

Definition 2

Farm behaviour in the given natural conditions is *ecologically optimal*, if the farm maximises its profit while respecting one or more sustainability indicators.

Mathematical model of farm ecological behaviour

To simulate the sustainability of Czech agriculture, a mathematical optimisation model of farm economic behaviour (FARMA-4) is applied, with the implementation of the above-mentioned ecological criteria in the sense of definition 2.

The adjusted mathematical model FARMA-4 includes the following segments:

- marketed and feeding commodities of the crop production on the arable land and on grassland;

- commodities of the livestock production connected with meat and milk outputs;

- feeding balance on the basis of self-supply in feedstuffs;

- calculations of production and income activities with respect to agricultural producer prices (farm-gate prices);

- cost calculations for all commodities on the basis of unit costs;

- calculations of commodity support (per hectare, per head or per production unit) on the basis of CAP rules or other defined policy scenarios;

- calculations of two optimisation criteria: 1) farm profit = total sales + total subsidies – total cost; and 2) farm profitability = (total sales + total subsidies) / total cost;

- calculations of the production of organic fertilisers (e.g. manure);

- calculations of the nutrient balance of NPK based on the circulation of fertilisers on the farm (industrial, organic (crop or animal in origin) and air-deposition) measured in the pure nutrients N, P and K;

- calculations of the total heads of animals measured by livestock units (LU) and LU/ha;

- yield calculations depending on the applied level of industrial fertilisers; and

- calculations of labour inputs depending on the production structure, measured by the total number of working hours or AWU (AWU = 2,200 hours/year).

According to the optimisation criterion (1 or 2) it is possible to compute farm profit maximisation in relation to additional conditions:

- positive nutrient balances of N, P, K;

- maximum or minimum LU on the agricultural land/forage land/grassland; and

- implementation of some agro-environmental programmes such as the maintenance of grassland.

For simulations of farm behaviour under the different production conditions of the Czech Republic, three farm categories were constructed for three regions (M is the area fully located in a non-LFA, P is the area located in a combination of non-LFA and LFA, and H is the area fully located in an LFA) and the two orientations/behaviours (P is the profit with the criterion 2 and I is the income with the criterion 1). All farm categories are represented as 100 ha farms where the structure of agricultural commodities and intensity parameters (hectare yields or milk yields) and cost parameters (unit costs per hectare or 'feeding days') are derived from the Czech farm surveys (FADN, CZ).

8.2.2   The multifunctionality indicators used

For modelling multifunctionality with the application of the model FARMA-4, the following indicators are used for the selected farm categories:

- the structure of land use, including arable land, grassland and land that has been set-aside (unused);

- the number of dairy/suckler cows;

- the livestock (ruminants) density (livestock units/ha);

- labour (employment); and

- the balance of elements (N, P, K).

8.2.3 Data

To simulate the impact of various policy scenarios, the following exogenous variables/parameters are applied:

- For the situation in 2004,
  - structure of production (using FADN data from 2004);
  - production costs for individual commodities (using the VUZE survey from 2004);
  - farm-gate prices for individual commodities (based on the report on the situation in Czech agriculture, 1994-2004); and
  - direct payments and LFA payments (using the database of policy measures for 2004, VUZE).

- For simulations of predictions related to the horizon of 2010,
  - direct payments (decoupled Single Payment Scheme) and LFA payments according to the Accession Treaty between the EU and the Czech Republic and according to the last known policy decisions/expectations;
  - farm-gate prices in the EU (OECD, 2005), reflecting reform in the sugar sector;
  - exchange rates CZK/EUR (VUZE predictions) and EUR/USD (OECD, 2005);
  - input prices, i.e. labour, land and other inputs (VUZE predictions); and
  - yields (VUZE predictions, based on AG-MEMOD simulations).

8.3 Policy scenarios

The policy scenarios below are applied for modelling the effects of policy measures on the multifunctionality of Czech agriculture.

- S1 represents the status quo for 2004–06, with decoupled SAPS payments, coupled top-up payments and LFA payments as in the period 2004–06.

- S2 incorporates full decoupling. It features the decoupled Single Payment Scheme at the maximum possible level and LFA payments based on the suppositions/conditions from the last draft of the Czech EAFRD (European Agricultural Fund for Rural Development) programmes,[5] sugar beet, milk and permanent crop productions ≤ the reference period 2004–06, and sugar beet production without compensation payments resulting from the reform.

- S3 models a reduced version of full decoupling, i.e. the same as the S2 scenario, but with a 20% reduction in direct payments.

- S4 also incorporates a reduced version of full decoupling, with increased agro-environmental support (the transfer of the financial sources from the first pillar to the second), i.e. the same as the S3 scenario, but with a 20% increase in payments for the agro-environmental scheme 'maintenance of pastures'.[6]

---

[5] For 2010 this is held to be a maximum of 24% of arable land on a farm in the H category, or 64% of arable land on a farm in the P category, and the ruminant density 0.36–1.8 LU/ha of forage land. The level of LFA payments is set at CZK 4,650/ha of grassland on a farm in the H category and CZK 3,410/ha of grassland on a farm in the P category. Investment support for the establishment of grassland is CZK 8,000 for all farm categories.

[6] In 2010 this is held to be a maximum of 23% of arable land on a farm in the H category, or 58% of arable land on a farm in the P category, and the ruminant density 0.36–1.0 LU/ha of forage land; a maximum of 170 kg N/ha on arable land and 40 kg N/ha of grassland. Compensation is set at CZK 3,100/ha of grassland for farms in the H category and CZK 2,800/ha for farms in the P and M categories.

The same conditions apply for investment support for the establishment of grassland and for the livestock density in the LFA payments, but the maximum share of arable land in farm acreage is 0.23% for the H farm category and 0.58% for the P farm category.

8.4 Results of modelling multifunctionality

The results of the simulations related to 2005 (S1) and to 2010 (S2–4) for the selected farm categories and the defined policy scenarios are presented in Table 8.5.

Table 8.5 Simulation results

Indicator/farm category	Unit	MI	MP	PI	PP	HI	HP
S1 (2005)							
Arable land	ha	99.5	84.0	71.4	12.9	52.6	8.2
– cereals & oilseeds	ha	76.7	63.6	64.9	6.0	47.0	3.1
– fodder	ha	2.7	2.3	4.5	6.8	5.6	5.0
– other arable land	ha	20.1	18.1	2.0	0.1	0.0	0.1
Grassland	ha	0.5	16.0	28.6	87.1	47.4	91.8
Unused land	ha	0.0	0.0	0.0	0.0	0.0	0.0
Dairy cows	heads	1.0	3.8	7.0	14.6	10.4	8.9
Suckler cows	heads	3.0	3.0	16.9	8.2	13.2	14.8
Livestock units/ha of fodder land	LU	2.55	0.81	1.50	0.53	0.94	0.52
Labour	AWU	1.47	1.50	1.78	1.55	1.70	1.34
Balance of N	kg	-11,580	-8,918	7,259	8,980	8,095	8,473
Balance of P	kg	-2,738	-2,478	73	200	-145	-56
Balance of K	kg	-4,165	-3,385	1,717	571	1,382	160
Operational surplus (profit)	000 CZK	1,239	1,281	1,220	1,122	1,236	1,036
S2 (2010)							
Arable land	ha	95.5	4.2	65.1	12.8	8.7	8.3
– cereals & oilseeds	ha	73.4	1.9	58.5	6.1	3.2	3.2
– fodder	ha	2.0	2.2	4.4	6.6	5.4	5.0
– other arable land	ha	20.1	0.1	2.2	0.1	0.1	0.1
Grassland	ha	4.5	31.0	34.9	87.2	91.3	91.7
Unused land	ha	0.0	64.8	0.0	0.0	0.0	0.0
Dairy cows	heads	1.0	3.8	7.0	14.6	10.4	8.9
Suckler cows	heads	3.0	3.0	16.9	8.2	13.2	14.8
Livestock units/ha of fodder land	LU	1.29	0.45	1.27	0.53	0.52	0.52
Labour	AWU	1.44	0.67	1.63	1.55	1.39	1.34
Balance of N	kg	-11,780	3,193	6,401	8,855	8,231	8,348

Table 8.5 cont.

Balance of P	kg	-2,817	92	-233	177	-70	-79
Balance of K	kg	-4,206	195	1,358	486	17	72
Operational surplus (profit)	000 CZK	751	436	1,219	1,424	1,410	1,352
S3 (2010)							
Arable land	ha	95.5	69.0	65.1	12.8	8.7	8.3
– cereals & oilseeds	ha	73.4	66.7	58.5	6.1	3.2	3.2
– fodder	ha	2.0	2.2	4.4	6.6	5.4	5.0
– other arable land	ha	20.1	0.1	2.2	0.1	0.1	0.1
Grassland	ha	4.5	31.0	34.9	87.2	91.3	91.7
Unused land	ha	0.0	0.0	0.0	0.0	0.0	0.0
Dairy cows	heads	1.0	3.8	7.0	14.6	10.4	8.9
Suckler cows	heads	3.0	3.0	16.9	8.2	13.2	14.8
Livestock units/ha of fodder land	LU	1.29	0.45	1.27	0.53	0.52	0.52
Labour	AWU	1.44	1.13	1.63	1.55	1.39	1.34
Balance of N	kg	-11,780	5,295	6,401	8,855	8,231	8,348
Balance of P	kg	-2,817	149	-233	177	-70	-79
Balance of K	kg	-4,206	-196	1358	486	17	72
Operational surplus (profit)	000 CZK	611	604	1,079	1,284	1,270	1,212
S4 (2010)+B23							
Arable land	ha	86.9	4.2	9.1	12.6	8.6	8.2
– cereals & oilseeds	ha	65.9	1.8	2.6	5.9	3.0	3.1
– fodder	ha	0.9	2.2	4.4	6.6	5.5	5.0
– other arable land	ha	20.1	0.2	2.1	0.1	0.1	0.1
Grassland	ha	13.1	31.0	90.9	87.4	91.4	91.8
Unused land	ha	0.0	64.8	0.0	0.0	0.0	0.0
Dairy cows	heads	1.0	3.8	7.0	14.6	10.4	8.9
Suckler cows	heads	3.0	3.0	16.9	8.2	13.2	14.8
Livestock units/ha of fodder land	LU	0.60	1.45	0.52	0.53	0.52	0.52
Labour	AWU	1.38	0.67	1.31	0.25	1.39	1.33
Balance of N	kg	-11,680	2,406	4,476	6,704	5,983	6,088
Balance of P	kg	-2,921	-64	-896	-267	-532	-540
Balance of K	kg	-4,180	115	-439	302	-171	-117
Operational surplus (profit)	000 CZK	640	382	1,276	1,522	1,555	1,499

Source: Authors' estimations.

The interpretation of the results for the 2005 (S1) and to 2010 (S2–4) scenarios can be summarised as below.

- S1 (status quo 2005) represents the optimisation of the current production structures under the 2004–06 policy conditions (not considering permanent crops). The optimum structures compared with the current ones show an increase in the acreage of arable land in all income-oriented farms and vice versa in all profit-oriented farms. Grassland can generate higher profitability; arable land can generate a higher amount of profit. Leaving the land unused is not an optimum solution for any of the farm categories.

- The results of S2–4 related to 2010 are very similar, substantiating the hypotheses on land use, labour inputs and so forth.

  - Profit-oriented farms could be more attracted by the conversion of arable land into grassland (even in non-LFA regions) and by the introduction of (relatively) very extensive cattle breeding, resulting in a reduction of labour inputs. Only in non-LFA regions could it be profitable to enlarge (to an extreme extent) the area of unused land.

  - Income-oriented farms could retain a higher share of arable land combined with relatively extensive cattle breeding, resulting in higher labour inputs. Gains in unused land may be very limited.

  - The decrease of direct payments by 20% (S3) would only impact the level of profitability and profits, without affecting land use or production structures.

- The implementation of higher payments for the maintenance of grassland (S4) could compensate the decrease of direct payments in the profitability and profits in all farm categories, but would be especially attractive for income-oriented farms in the potato regions.

Conclusions – Policy issues and recommendations

Around 75% of the utilised agricultural area in the Czech Republic is occupied by profit-oriented farms. Because of a combination of expectations about the slow progress in land consolidation (re-parcelling), a zero-level of degressivity by the size of farm in direct payments until 2010, a low level of degressivity by the size of farm in LFA payments and other factors, the share of profit-oriented farms in the Czech Republic may even rise.

With this kind of farm structure in mind, the total decoupling of direct payments combined with the possibility that land will not have to be used for the production of a commodity [7] can lead to an extreme level of extensive farming and a large share of unused land in the most productive regions.

These trends, however, could be counter-balanced by the conditions for LFA payments, which contribute significantly to the finances of farms in the LFAs. On the one hand, the maximum limits for the share of arable land and on the other hand the minimum limits for livestock density could lead to an enlargement of the grassland acreage accompanied by a shift of cattle breeding on the LFA land.

Under these conditions, the tendency towards the intensity of production along with a reduction in labour costs and in employment by farms could be smoothed. This point applies to all regions and farm categories. Nevertheless, there may be a perpetual risk that the expected positive externalities from farming related to the environment (water, soil) may be eliminated by the negative externalities connected with rural employment or to rural social and human capital.

Taking into account all aspects of multifunctionality, the main policy issues and recommendations deduced from the model simulations to be addressed by policy-makers are

- Implementation of a graduated scale for degressivity in the direct payments and particularly in the LFA payments could create stimuli for the development of small and medium-sized farms, generating better conditions for job opportunities and an increase in the quality of human and social capital in (marginal) rural areas.

- Greater support is needed for the development of non-agricultural activities on farms or for the development of micro-firms in rural areas.

- Agro-environmental schemes need to be implemented, based on more stringent conditions for compensation payments or on schemes to compensate non-commodity outputs in accordance with real environmental effects (to reduce the risk that agro-environmental

---

[7] The land will however, have to be maintained according to good farming practices/cross-compliance.

payments, owing to low transaction costs in the required changes to farm practices, are considered a prolongation of direct payments).

- In any case, decoupling can lead to a fall in the volume of production by the Czech agricultural sector and contribute to a reduction of surpluses under the EU-25 framework.

Bibliography

Doucha, T. and E. Divila (2004), "Possible impacts of the Czech agricultural policy after the EU accession on the land market and land usage", *Agricultural Economics*, Vol. 51, No. 5, pp. 185-93.

Doucha, T., E. Divila and M. Fischer (2005), "Land use and ownership and the Czech farm development", in *Rural Areas and Development*, Vol. 3, European Rural Development Network, Institute of Agricultural and Food Economics, Institute of Geography and Spatial Organization, Polish Academy of Sciences, Warsaw.

Dwyer, J., D. Baldock, J. Wilkin and D. Klepacka (2005), *Scenarios for modelling trade impacts upon multifunctionality in European agriculture*, ENARPRI Working Paper No. 10, CEPS, Brussels, January.

Foltýn, I., T. Zídek and I. Zedníčková (2004), "Impacts of accession of the Czech Republic to the EU on the sustainable development of the Czech agriculture", presentation for the conference held by the Lithuanian Institute of Agricultural Economics, Vilnius, November.

Foltýn, I., I. Zedníčková and L. Grega (2002), "Efficiency of agricultural policy as a factor of sustainability of protected areas", in I. Camarda, M.J. Manfredo, F. Mulas and T.L. Teel (eds), *Global Challenges of Parks and Protected Area Management, Proceedings of the 9th ISSRM*, La Maddalena, Sardinia, ISBN 88-7138-318-4.

Guyomard, H. (2006), *Review of methodological challenges: Multifunctionality and trade*, ENARPRI Working Paper No. 7, CEPS, Brussels.

Ministerstvo zemědělství ČR [The Czech Ministry of Agriculture] (1994-2004) *Zprávy o stavu zemědělství ČR, 1994–2004* [Reports on the situation in the Czech agriculture, 1994–2004], Ministerstvo zemědělství ČR, VÚZE, Prague.

OECD (2002), *Multifunctionality: Towards an Analytical Framework*, OECD, Paris.

————— (2005), *OECD-FAO Agricultural Outlook 2005-2014*, OECD, Paris.

9. The impact of potential WTO trade reform on greenhouse gas and ammonia emissions from agriculture
A case study of Ireland

Trevor Donnellan and Kevin Hanrahan

Introduction

This study combines an economic, partial-equilibrium, agricultural commodity and inputs model (the FAPRI-Ireland model) [1] with a model for the estimation of greenhouse gases and ammonia emissions from agriculture. It considers a potential reform of agricultural trade policy under a possible World Trade Organisation (WTO) agreement, to reveal the extent to which there are environmental effects associated with such a reform that need to be considered in addition to the conventional economic ones.

Since the industrial revolution the use of fossil fuels (oil, coal and natural gas) has provided power for industry and facilitated the lifestyle of Western societies. Owing to the use of fossil fuels, levels of atmospheric carbon dioxide have risen, which may augment the greenhouse effect to the point where a change in the climate may result. Higher levels of other trace gases such as nitrous oxide (N_2O), methane (CH_4) and chlorofluorocarbons (CFCs) may also contribute to a change in climatic conditions.

---

[1] The FAPRI-Ireland Partnership is a research affiliation between Teagasc – The Irish Agriculture and Food Development Authority and the Food and Agricultural Policy Research Institute (FAPRI) based at the University of Missouri (for further information see http://www.tnet.teagasc.ie/fapri).

Collectively these gases are referred to as greenhouse gases (GHGs). In Ireland agricultural production is a leading contributor of GHG emissions to the atmosphere in the form of methane and nitrous oxide.

While some remain sceptical about the evidence of global warming, the body of scientific opinion contends that a significant alteration of our climate is possible within this century. Continuing global warming may affect, among other things, crop yields and water supply. Furthermore, it may generate the potential for altering the range and number of pests that affect plants as well as diseases that threaten the health of both humans and animals. An increase in global temperatures may cause the melting of polar icecaps, which would raise sea levels and inundate low-lying land areas around the world.

Reflecting growing international concern about global warming, the Kyoto Protocol[2] was signed in Japan in 1997. It resulted in specific limitations for GHG emission levels to be achieved by the first commitment period 2008–12 in countries that are signatories to the agreement. These targets were set with reference to GHG levels in 1990. Most developed countries must reduce their GHG emissions below the 1990 level to comply with the Protocol. Within the EU, Ireland received a concession that allows an increase in its GHG emissions by no more than 13% above the 1990 levels by the first commitment period.

In 2000 the National Climate Change Strategy (NCCS) for Ireland was published. It projected that without policies to contain the level of emissions, Ireland would in fact exceed its target of 60.74 million tonnes (Mt) of carbon dioxide (CO_2) equivalent by up to 22% by the first commitment period.[3] In the NCCS, the Department of the Environment, Heritage and Local Government in Ireland set out specific measures to control GHG emissions.

Relative to other EU member states and most other developed countries, Ireland is unusual in terms of the percentage contribution made by agriculture to national GHG emissions. Of the 68 Mt of GHG CO_2 equivalent produced in Ireland in 2004, it is estimated that 28% was

---

[2] See the US Department of State, Bureau of Oceans and International Environmental and Scientific Affairs (1998) for more details.

[3] See the NCCS, Department of the Environment (2000), p. 12.

contributed by Irish agriculture (Environmental Protection Agency, 2005). This figure reflects both the high degree of agricultural activity and relatively lower levels of other GHG sources (such as heavy industry) in Ireland. The emission of GHGs from Irish agriculture principally comes from animals but is also the result of agricultural practices such as the use of fertiliser and manure management. It is likely that policy-makers will seek to reduce GHG emissions below the levels projected in the NCCS report. In this regard they may consider the cost of reducing emissions from each sector in order to minimise the effect on the overall economy. There is thus a need to estimate GHG emissions from the various sectors of the economy, including agriculture.

This study projects the future level of GHG emissions under existing agricultural policies prevailing in the EU and then contrasts that outcome with projections made under an assumed WTO agreement, thereby capturing the potential impact of such a trade reform for GHG emissions from Irish agriculture.

In addition to concerns relating to GHG emissions, since the 1970s there has been growing international concern about air pollution. In the EU an objective of policy-makers is to formulate and implement strategies to improve air and water quality. To meet this objective, the control of emissions from a variety of industrial, commercial and agricultural sources is a key aim. With this in mind the European Council issued a Directive (No. 2001/81/EC) in 2001 that sets limits for each EU member state in terms of total emissions of specific gases. These limits are to be met by 2010.

Four categories of pollutants – sulphur (SO_2), nitrous oxides (NO_x), volatile organic compounds (VOCs) and ammonia (NH_3) – have been identified as being responsible for acidification and eutrophication of ground water and ground-level ozone pollution. The Directive allows EU member states to provide their own mechanisms to ensure that reduction targets are achieved. As part of the Directive member states will be required to report each year on their actual and projected future levels of emissions of these substances. National programmes are required to specify how national ceilings will be met. The Directive contained provisions for reviews in 2004 and 2008 to identify the progress being made and whether further actions are required.

Some of the pollutants mentioned above can be transported considerable distances through the air or in water, which means that pollution arising in one country may have an impact in another. Thus a coordinated international approach, which extends beyond the EU, is required to address the issue. Accordingly, in November 1999 EU member states together with Central and Eastern European countries, the US and Canada negotiated the UNECE Gothenburg Protocol to the 1979 Convention on Long Range Transboundary Air Pollution to Abate Acidification, Eutrophication and Ground Level Ozone (UNECE, 1999). The Gothenburg Protocol contains emission ceilings that are not as stringent as are those agreed by the European Council. Under the Gothenburg Protocol, Ireland agreed to reduce its NH_3 emission levels by 9% from those estimated for 1990. With regard to Irish agriculture's contribution to these forms of pollution, a number of consequences can be identified as below.

Eutrophication refers to the gradual increase in the concentration of phosphorus, nitrogen, ammonia and other plant nutrients in water ecosystems such as lakes. As the amount of organic material that can be broken down into nutrients rises, the productivity or fertility of such an ecosystem increases. Runoff from land may enter water systems containing, among other things, fertiliser and decomposing plant matter. This spillover can cause algal blooms (highly concentrated amounts of micro-organisms) to develop on the water surface, which then prevents the light penetration and oxygen absorption that is necessary for aquatic life. This process can be intensified when excessive amounts of fertilisers (as well as sewage and detergents) are prevalent. Ammonia is a major constituent of agricultural fertilisers, which contributes to the process of eutrophication.

Acidification can result from emissions of sulphur dioxide, nitrogen oxides and ammonia. Although sulphur is the biggest contributor to acidification, nitrogen compounds are also a significant source. When soil becomes acidified it can cause nutrients to leach, which then reduces soil fertility. Metals can also be released from the process, which can affect the micro-organisms that facilitate the decomposition of organic matter in the soil and in turn affect birds, animals and humans. Tree damage such as leaf and needle losses has been linked to acidification and high concentrations of ground ozone.

We examine the level of ammonia produced by the various sub-sectors of Irish agriculture. We use economic projections for future levels of agricultural activity in conjunction with per unit estimates of ammonia emissions to calculate future levels of ammonia emissions from Irish agriculture.

In this study we do not consider the issue of whether or not GHG or ammonia emissions from agriculture should be considered as a multifunctional output of the agricultural sector. The OECD has produced an analytical framework wherein the nature and definition of multifunctionality is discussed at length (OECD, 2001).

The rest of this chapter is divided into four further sections. Section 9.1 examines the methodology for the estimation of the impact of trade policy on the level of agricultural activity and in turn the effects of GHG and ammonia production. Section 9.2 outlines two states of the world for examination. The first, referred to as a baseline, examines agricultural activity and emissions generation under a continuation of existing (Uruguay round) WTO trade policies and the current (Luxembourg Agreement) EU common agricultural policy (CAP). The second state of the world, a WTO reform scenario, alters trade policies (as a result of a hypothetical WTO agreement) to assess the impact on agricultural activity and emissions generation. The policy change considered under the WTO reform is also detailed in this section. The difference between emission levels under the two scenarios is an estimate of the environmental effects of the WTO reform. Section 9.3 presents the results for agricultural production, GHG and ammonia emissions under both the baseline and the WTO reform scenarios. The results are followed by some conclusions and areas for further work.

9.1   Method of analysis

The approach used here involves the use of two distinct modelling frameworks, which interact with each other to produce projections of the impact of trade policy reform on GHG and ammonia emissions. The first component is an econometric, partial-equilibrium commodity model and the second component is the satellite emissions projection models for both GHG and ammonia.

9.1.1   Partial-equilibrium commodity model

The FAPRI-Ireland model is a set of econometric, dynamic, multi-product, partial-equilibrium commodity models. In its current version, the model has an agricultural commodity coverage that extends to markets for grains (wheat, barley and oats), other field crops (potatoes, sugar beet and vegetables), livestock (cattle, pigs, poultry and sheep) and milk and dairy products (cheese, butter, whole milk powder and skim milk powder).

Many of the equations in the model are estimated using annual data from the period 1973–2005 or over shorter periods in cases where data are not available or where, for policy reasons, longer estimation periods would not be meaningful.

The FAPRI-Ireland model is structured as a component of the FAPRI EU GOLD model, which is a commodity model of EU agriculture. The GOLD model in turn can form a component of the FAPRI world modelling system for world agriculture. In this way the model for Ireland can incorporate the consequences of changes in international trade policy as they relate to agriculture.

The primary purpose of the FAPRI-Ireland model is to analyse the effect of policy changes on economic indicators such as the supply and use of agricultural products, agricultural input expenditure and sector income. In so doing the model produces future projections of animal numbers, input usage volumes (e.g. fertiliser, feed, fuel and energy) and other indicators. These data can be incorporated into the satellite GHG models to enable the provision of base data and projections relating to multifunctionality indicators, such as GHG emissions, fertiliser usage and ammonia emissions. Key components of the structure of the partial-equilibrium model are set out below.

The equation for the total agricultural area farmed is modelled as:

$$taf_t = f\left(agout_{t-1} \Big/ gdp_{t-1} \right) \tag{1}$$

where taf_t is the total agricultural area in year t and $agout_{t-1}$ is the value of agricultural output in year $t-1$ and gdp_{t-1} is a measure of national income in year $t-1$. The equations used to determine the share of the total agricultural area farmed within each agricultural culture group can be expressed as:

$$ash_{i,t} = f\left(ret_{i,t-1}, agout_{t-1}, ash_{i,t-1}, V_t, Z_t\right) \quad i = 1,\ldots,5 \tag{2}$$

where $ash_{i,t}$ is the share of the total agricultural area to be allocated to i-th culture group in year t, $ret_{i,t-1}$ is the value of the output from the i-th culture group and $agout_{t-1}$ is the value of total agricultural output in year $t-1$, while V and Z are vectors of exogenous and endogenous variables that could have an impact on the area allocated to agriculture culture group i. The land use associated with one of the five agriculture culture groups modelled (pasture, hay and silage, potatoes, sugar beet and cereals) is derived as the residual land use so as to ensure land-use balance.

The total area allocated to the i-th agricultural culture group is then derived as the product of the i-th area share times the total agricultural area:

$$af_{i,t} \equiv ash_{i,t} * taf_t \tag{3}$$

Within each of the i agricultural culture groups, land may be further allocated among competing cultures, for example within the land area allocated to the cereals culture group soft wheat 'competes' with barley and oats for land. Within the culture group allocation of land this is modelled using area allocation equations of a similar form to (2):

$$asf_{i,t}^j = f\left(ret_{i,t-1}^j, \sum_{\substack{k=1 \\ k \neq j}}^{m} ret_{i,t-1}^k, asf_{i,t-1}^j, S_t, W_t \right) \quad j,k=1,\ldots,m \tag{4}$$

where $asf_{i,t}^j$ is the share of the j-th culture within the culture group i, $ret_{i,t-1}^j$ is the return to the j-th culture in year $t-1$, and S_t and W_t are other endogenous and exogenous variables that may affect the allocation of land among the j competing cultures within any given culture group i. The land (in hectares) allocated to the j-th culture is then derived as the product of the total land allocated to the i-th culture group ($af_{i,t}$) times the area share ($asf_{i,j,t}$):

$$aha_{i,t}^j \equiv asf_{i,t}^j * af_{i,t} \tag{5}$$

The yield equations of culture k in culture group i can be written as:

$$r_{i,t}^k = f\left(p_{i,t-1}^j, r_{i,t-1}^k, V\right) \quad j,k=1,\ldots,n \tag{6}$$

where $r_{i,t}^k$ is the yield per hectare of culture k belonging to the culture group i, and V is a vector of variables, which could influence the yield per hectare of the culture being modelled.

On the demand side, crush and feed demand and non-feed use per capita are modelled using the following general functional forms:

$$Fu_{i,t}^k = f\left(p_{i,t}^j, Z\right) \quad j,k=1,\ldots,n \tag{7}$$

where $Fu_{i,t}^{k}$ is the feed demand for culture k belonging to the culture group i and Z is a vector of endogenous variables (such as the level of meat production), which could affect the feed demand;

$$NFu_{i,t}^{k} = f\left(p_{i,t}^{j}, NFu_{i,t-1}^{k}\right) \quad j,k = 1,...,n \tag{8}$$

where $NFu_{i,t}^{k}$ is the non-feed demand for culture k belonging to the culture group i and V is a vector of exogenous variables (such as income) that could have an impact on non feed demand;

$$CR_{i,t}^{k} = f\left(p_{i,t-1}^{h}, p_{i,t-1}^{h}, p_{i,t-1}^{l}, CR_{i,t-1}^{h}\right) \quad h, l = 1,...,n \tag{9}$$

where $CR_{i,t}^{k}$ is the crush demand for oilseed culture k and $p_{i,t-1}^{h}$ is the real price of considered seed oil and $p_{i,t-1}^{l}$ is the real price of the seed meal produced as a product of the crushing process.

While the structure of individual livestock sub-models varies, their general structure is similar and is presented below. Ending numbers of breeding animals can be written as:

$$cct_{i,t} = f\left(cct_{i,t-1}, p_{i,t}, V\right) \quad i = 1,...,n \tag{10}$$

where $cct_{i,t}$ is the ending number in year t for the breeding animal type i, $p_{i,t-1}$ is the real price in year $t-1$ of the animal culture i considered, and V is a vector of exogenous variables that could have an impact on the ending inventory concerned (such as the direct payment linked to the animals concerned or specific national policy instruments).

Numbers of animals produced by the breeding herd inventory can be written as:

$$spr_{i,t} = f\left(cct_{i,t-1}, ypa_{i,t}\right) \quad i = 1,...,n \tag{11}$$

where $spr_{i,t}$ is the number of animals produced from breeding herd $cct_{i,t}$ in year t and $ypa_{i,t}$ is the exogenous yield per breeding animal concerned.

Within each animal culture i there may be m categories of slaughter j. The number of animals in animal culture i that are slaughtered in slaughter category j can be written as:

$$ktt_{i,t}^{j} = f\left(cct_{i,t}^{j}, p_{i,t}, z_{i,t}^{j}, V\right) \quad i = 1,...,n \quad j = 1,...,m \tag{12}$$

where $ktt_{i,t}^{j}$ is the number of animals slaughtered in category j of animal culture i in year t, $z_{i,t}^{j}$ is an endogenous variable that represents the share of different categories of animals slaughtered in the total number of animals slaughtered for the animal culture concerned, and V is a vector of exogenous variables.

Ending stocks of animals (breeding and non-breeding) are derived using identities involving initial inventories of animals, animal production (births), slaughter, and live exports and imports.

The number of dairy cows can be written as:

$$cct_t = f(p_t, V) \tag{13}$$

where cct_t is the ending number of dairy cows in year t, p_t is the real price of milk in year t, and V is a vector of exogenous variables that could have an impact on the ending inventory concerned (including policy instruments such as the milk quota).

Milk yields per cow can be written as:

$$r_t = f(p_t, V) \tag{14}$$

where r_t is the milk yield per cow, p_t is the real price of milk in year t, and V is a vector of variables that could influence the yield per cow.

9.1.2 GHG emissions model

The projections of commodity outputs and input usage from the FAPRI-Ireland model can be converted into projections of GHG emissions using the default conversion coefficients outlined by Houghton et al. (1996) in their contribution to the Intergovernmental Panel on Climate Change (IPCC), modified, where possible, with specific coefficients for Ireland.

The methodology for the establishment of the GHG inventories was proposed by Houghton et al. (1996). It was subsequently adopted and adjusted to allow for conditions specific to Ireland by the Department of the Environment (1997). The approach essentially involves applying conversion coefficients to agricultural data and calculating the associated emissions of GHGs from enteric fermentation, manure management practices and agricultural soil management as defined by Houghton et al. (1996).

Data on Irish livestock numbers, enterprise areas and input applications have been obtained from the FAPRI-Ireland model. Livestock emission factors are expressed in terms of the annual amount of methane produced by the animal. These emission factors vary by animal type, not only because of their differing size and feed consumption, but also because of the manner in which food is digested and the animal manure is subsequently treated.

Concerning manure management, the nature of production systems tends to favour the management of cattle and pig manure in liquid systems, which facilitate anaerobic respiration and the emission of methane. By contrast, sheep are rarely housed and consequently methane emissions from their manure are negligible.

The emission of GHGs from agricultural soils varies in accordance with the manner in which the land is managed, which in turn depends on the type of crop production system in place. For the purposes of emissions calculations, the IPCC categorises farmland under three uses. Crop land and more intensively farmed grassland have quantities of fertiliser applied to them, whereas less intensively farmed grassland may have no fertiliser applied to it. Consequently, the levels of methane and nitrous oxide emissions from cropland and more intensively farmed grassland are considerably higher than grassland maintained without fertiliser.

GHGs in the form of methane and nitrous oxide emissions from each agricultural sub-sector i are thus a function of the number of animals, crop areas harvested and nitrogen application. Since the global warming potential of CH_4 and N_2O differ, for the purpose of their addition these are brought to a common base of CO_2 equivalents using standard weighting systems. CH_4 produced in each agricultural sector can be represented as:

$$CH_{4,i,t} = f\left(q_{i,t}, \alpha_i\right)$$
(13)

where $CH_{4,i,t}$ is the total amount of CH_4 produced by sector i in year t, q is the quantity of animal or crop category i in year t and α is the methane conversion coefficient associated with the animal or crop category i.

Similarly, N_2O produced in each agricultural sector can be represented as:

$$N_2O_{j,t} = f\left(q_{j,t}, \beta_j\right)$$
(14)

where $N_2O_{j,t}$ is the total amount of N_2O produced by sector j in year t, q is the quantity of animal or crop category j in year t and β is the nitrous oxide conversion coefficient associated with the animal or crop category j.

Finally, total GHG emissions in the common base of CO_2 equivalents can be expressed as:

$$GHG_t = \delta \sum_{i=1}^{n} CH_{4,i,t} + \gamma \sum_{j=1}^{m} N_2O_{j,t}$$

(15)

where EquivCO$_2$ is CO_2 equivalent, while $\delta = 21$ and $\gamma = 310$ are the global warming potentials of methane and nitrous oxide respectively.

The next section provides a brief review of the results for the agricultural variables used in the generation of GHG emissions. Then the consequent baseline and alternative scenario projections of GHG emissions from Irish agriculture are presented.

9.1.3 Ammonia emissions model

The projections of commodity outputs and input usage from the FAPRI-Ireland model can be converted into projections of ammonia emissions using conversion coefficients. Estimates are based on the quantities of synthetic fertiliser applied (e.g. urea and calcium ammonium nitrate) per hectare. From 2000 onwards projections are calculated to allow for lower nitrogen (N) application rates on the land areas that participate in the Rural Environment Protection Scheme. This scheme is an income-support measure for farmers in Ireland, which promotes farming practices that allow environmentally-friendly food production and the conservation of wildlife habitats.

9.2   Descriptions of the baseline and WTO reform scenarios

The method of reporting the effect of trade policy on GHG and ammonia emissions relies upon a comparison of two states of the world, one including, and the other excluding, the trade policy change under examination.

Baseline scenario. This scenario calculates the level of activity that would arise in the future under a base case set of agricultural policies. Projections of activity levels under the base case of agricultural policy are referred to as the baseline policy outcome.

The baseline projections of agricultural activity used in this section are drawn from the baseline outlined in Binfield et al. (2006), i.e. the CAP mid-term review and the GATT[4] Uruguay Round Agreement on Agriculture (URAA).[5] Projections of GHG emissions stemming from these agricultural projections are presented below.

Alternative scenario. This scenario calculates the level of activity that would arise in the future under alternative agricultural policies. Projections of activity levels under alternative policies are referred to as the alternate policy outcomes.

At the time of writing (December 2005), the outcome of the WTO Doha round negotiations is unknown. The WTO reform scenario formulated and analysed here is close to the current position of the EU within the Doha round (EU Trade Commissioner Peter Mandelson's offer of 28 October 2005). Under the WTO scenario, as defined in Table 9.1, the aggregate measure of support (AMS) is cut by 70% from the bound URAA levels. Under the export competition heading, the EU phases out its export subsidies over the course of 10 years. Also, in this WTO scenario, 50% of the cut in export subsidies is front-loaded on the first year (2007) with the remaining 50% phased out in equal instalments over the following nine years. Under the market access headings a cut in average tariffs of 60% is implemented with lower cuts in tariffs applying to sensitive products set at 25%. Beef and butter are designated as sensitive products for the EU and are subject to these lower tariff reductions. No other market access provisions (e.g. tariff-rate quotas or TRQs) are altered.

Under the WTO reform scenario analysed, the green and blue box classifications of current government support to agriculture are retained and unaffected by the changes proposed.

---

[4] GATT refers to the General Agreement on Tariffs and Trade – the precursor to the World Trade Organisation.

[5] Note that the more recent reform of the EU sugar regime (Council Regulation (EC) No. 318/2006) is not reflected in this analysis.

Table 9.1 WTO reform scenario

	Domestic support	Export subsidies	Market access
WTO scenario	70% reduction in the total AMS based on Uruguay round final bound levels with retention of green and blue boxes	Phased out over 10 years, with a 50% down payment in year 1 and 9 years of equal instalments thereafter	60% average cut in tariff lines, with a 25% minimum cut (to apply to products designated as sensitive)

Source: Authors' compilation.

The effect of the change in policy can be measured by the difference between the projections for the baseline and the WTO reform scenarios.

9.3   Results

Here the results for GHG emissions are presented under both the baseline policy and the WTO reform scenario. The results include a summary of the impact on agricultural production levels as well as details on GHG emissions.

9.3.1   *Irish agricultural production: Baseline policy*

Under baseline policies, livestock numbers in Ireland are projected to fall over time. The number of dairy cows would fall as a result of the quota limits on total milk production and genetic improvements that lead to dairy cows becoming more productive over time – thus the number of cows required to fill the quota would decrease. Dairy cows are by far the largest source of agricultural GHG emissions, on a per head basis, so this reduction would have a sizable effect on total Irish agricultural GHG emissions.

Under the baseline projection, the decoupling of agricultural policy as recently introduced in the EU will lead to a decrease in beef cattle and sheep numbers over the period to 2015, since the policy will make it unprofitable for some producers to raise these animals. In Ireland, the baseline number of pigs and other animal categories is projected to remain relatively static over the projection period.

The total land area in agricultural use in Ireland will have declined slightly by about 1% under baseline policies by 2015 relative to the level in 2004. Some changes in land use are projected over the period, as there is a slight tendency for area planted with cereals and root crops to shift into use as pasture. Although animal numbers are expected to decline, the move towards more extensive livestock production will mean that the proportion of land devoted to pasture, hay and silage will not change markedly. Conditions attached to the receipt of the decoupled payments limits the extent to which land will move between these use categories.

9.3.2 GHG emissions from Irish agriculture: Baseline policy

The baseline projections for total emissions from agriculture are presented in Figure 9.1. Overall, the baseline projections suggest that, with the introduction of decoupling as an agricultural policy, there will be a reduction in overall agricultural activity. Consequently, Irish agricultural GHG emissions are also set to decline. The reduction comes mainly through a decrease in the projected future numbers of cattle (both dairy and beef) and sheep. Total GHG emissions from Irish agriculture are projected to fall by approximately 14% by 2015 relative to 2004. Measured against a 1990 base, the decline by 2015 is projected to be over 16%.

Emissions must be reduced by 8% for the EU-15 as a whole, by the first commitment period. Yet under the EU Burden-Sharing Agreement, Ireland is committed to minimising its rate of increase in GHG emissions to 13% above the 1990 level agreed under the terms of the Kyoto Protocol. Strong economic growth has prompted a significant rise in emissions in the non-agricultural sectors of the Irish economy since 1990, so the projected fall in agricultural GHG emissions would represent an important contribution towards the attainment of Ireland's GHG emissions target. Projected emissions under the baseline scenario are shown in Table 9.2.

Figure 9.1 Projections of GHG emissions from Irish agriculture – Baseline policy

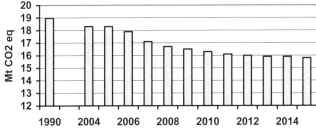

Note: Totals represent CH_4 and N_2O (in CO_2 equivalents) from enteric fermentation, manure management and agricultural soils.

Source: FAPRI-Ireland Partnership Model (2006).

Table 9.2 GHG emissions by Irish agriculture from 1990 to 2015 – Baseline policy

Source category	Unit[a]	1990	Baseline 2015	Change (%)
Methane (CH$_4$)	Gg	551.6	469.8	-14.83
Nitrous oxide (N$_2$O)	Gg	23.9	21.3	-10.88
Total (CO$_2$ equivalent)	Mt	18.97	15.8	-16.71

[a] Gg = gigagram (thousand tonnes); Mt = million tonnes

Note: The C0$_2$ equivalent measure represents the change in the global warming potential of methane and nitrous oxide.

Source: FAPRI-Ireland Partnership Model (2006).

The next section outlines the results of the WTO reform scenario using the FAPRI-Ireland model. The consequent effects on GHG emissions under these alternate policy scenarios are presented.

9.3.3   Projections of agricultural activity: WTO reform scenario

It is projected that under the WTO reform scenario, milk quotas will continue to be filled in Ireland. Dairy cow numbers will decline at a slightly lower rate than indicated in the baseline, because the WTO reform scenario will lead to a reduction in milk prices that is greater than in the baseline. This outcome slightly impedes the growth in milk yields; as a corollary, it also slows the fall in cattle numbers.

Under the WTO scenario, there is also a further contraction in Irish beef cattle numbers as reduced exports (due to the elimination of export refunds) and increased imports (due to reduced import tariffs) lead to lower beef prices across EU member states including Ireland. Overall, cattle numbers under the WTO scenario are lower than in the baseline.

In the case of sheep, Irish prices and production also decline as imports from outside the EU increase (due to lower import tariffs). The number of pigs and other animal categories is projected to remain relatively static over the projection period under the WTO reform scenario.

Relative to the baseline, the WTO reform scenario leads to only minor changes in the allocation of Irish farmland to pasture, hay and silage, cereals and root crops. As indicated under the baseline projections, the conditions attached to the receipt of the decoupled payments will limit the extent to which land will move between use categories.

9.3.4 Projections for GHG emissions: WTO reform scenario

Since the WTO reform scenario suggests that cattle and sheep numbers fall appreciably relative to the baseline levels, methane emissions from both enteric fermentation and manure management are expected to decline by a greater extent in this scenario than under the baseline.

Emissions levels under the WTO reform scenario for methane, nitrous oxide and GHG equivalent emissions of CO_2 are illustrated in Table 9.3. Under the WTO reform scenario, by 2015 the total GHG emissions from agriculture are expected to decrease by 3.5 Mt of CO_2 equivalent (a decrease of almost 20%) relative to the position in 1990.

Table 9.3 GHG emissions by Irish agriculture from 1990 to 2015 – Luxembourg Agreement/EU WTO scenario

Source category	Unit[a]	1990 Actual	2015 WTO reform scenario	Change (%)
Methane (CH_4)	Gg	551.6	456.9	-17.17
Nitrous oxide (N_2O)	Gg	23.9	20.9	-12.55
Total (CO_2 equivalent)*	Mt	18.97	15.3	-19.35

[a] Gg = gigagram (thousand tonnes); Mt = million tonnes

Note: The CO_2 equivalent measure represents the change in the global warming potential of Methane and Nitrous Oxide.

Source: FAPRI-Ireland Partnership Model (2006).

By contrast, the baseline analysis presented earlier projected a reduction of 3.2 Mt of CO_2 equivalent relative to the 1990 level (a decrease of over 16%). Under the WTO reform scenario the 2015 outcome represents a reduction in emissions relative to 1990 levels that is almost 3 percentage points below the reduction projected to occur in the baseline. This result suggests that the WTO reform examined here would deliver additional environmental benefits to those already anticipated under the baseline agricultural reforms taking place in the EU.

Figure 9.2 presents the projections for GHG emissions from Irish agriculture under the baseline and the WTO reform scenarios in CO_2 equivalent terms.

Figure 9.2 Projections of GHG emissions from Irish agriculture – Baseline and WTO scenarios

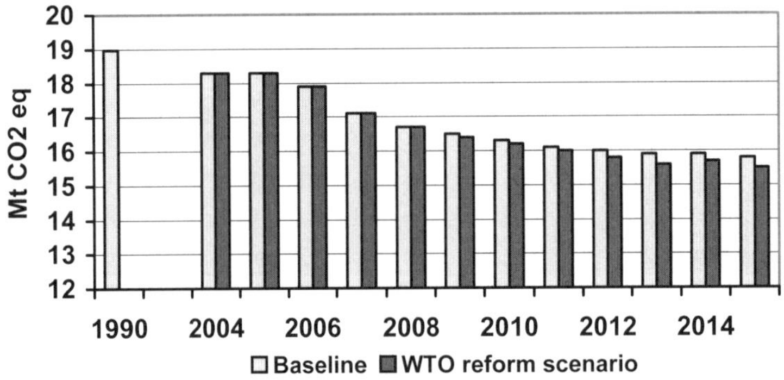

Note: Totals represent CH_4 and N_2O (in CO_2 equivalent) from enteric fermentation, manure management and agricultural soils.

Source: FAPRI-Ireland Partnership Model (2006).

9.3.5 Projections of ammonia emissions: Baseline and WTO reform scenarios

Apart from any environmental restrictions that might come into place, the type of agricultural policy pursued in the future will affect the level of agricultural activity and in turn the total level of ammonia emissions. The level of ammonia emissions can be projected under the baseline and WTO scenarios. The approach builds on earlier work (Behan & Hyde, 2003).

It is found that under both the baseline and the WTO scenarios emissions of ammonia are likely to decline relative to current levels. By 2015 it is projected that ammonia emissions from agriculture will have declined by 13% relative to the 2004 level. Despite the decrease this would still mean that the level of ammonia emissions from agriculture would be very close to the 1990 base year for the Gothenburg Protocol. Ireland's commitment under this Protocol is for a 9% reduction on the 1990 level of emissions of ammonia (a target level of 116,000 tonnes) in aggregate economy-wide terms by 2010. Projections of ammonia emissions are shown in Figure 9.3.

The reduction in emissions in the baseline stems from the decoupling of payments, which results in fewer beef cattle and sheep numbers. With milk production fixed by a quota, the continuing increase in milk yields per cow means that dairy cattle numbers are also reduced in the baseline.

Figure 9.3 Projected ammonia emissions from agriculture – Baseline and WTO scenarios

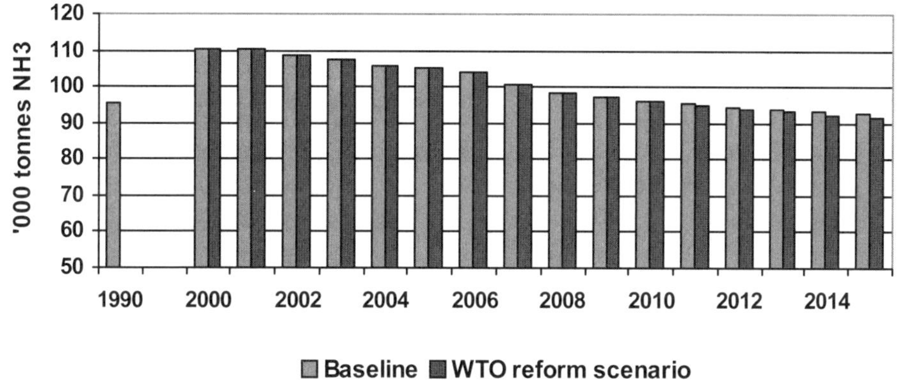

☐ Baseline ■ WTO reform scenario

Source: FAPRI-Ireland Partnership Model (2006).

Under the WTO reform scenario it is projected that there will be further reductions in beef cattle and sheep numbers relative to the levels projected in the baseline but the decline in dairy cattle numbers is in line with that of the baseline. Relative to 2004 the reduction achieved by 2015 represents a decline of 14%. The extent of the decline is only slightly (1%) greater under the WTO scenario compared with the baseline.

The analysis here suggests that the effects on non-dairy sectors of EU and Irish agriculture of the WTO elements of the scenario analysed would be somewhat modest. The changes that arise under the scenario relative to the baseline in these sectors largely stem from policy changes in the Luxembourg Agreement. Nevertheless, more extensive trade reforms might have a more widespread impact on agriculture in the EU and Ireland. Results will also be sensitive to the future exchange rate between the euro and the US dollar.

9.3.6 Comments on the overall results

The overall results projected for both GHG and ammonia emissions suggest that the reductions in emissions foreseen over time will largely arise from CAP reforms rather from international trade policy (WTO) reforms. Although this is the most obvious conclusion to make, it may also be slightly misleading. The motivations for reform of the CAP relate, to some degree, to pressures external to the EU – principally the need to make the CAP more compatible with a future WTO agreement. It is unlikely that the 2003 CAP reform would have taken the shape it did, had it not been for these WTO-related pressures. Therefore one could argue that the reductions in emissions projected under the baseline are also motivated by trade policy reform and not merely by changes to domestic policies.

Conclusions

This study projects some of the effects of recent reforms to EU agricultural policy (as a baseline) on the environmental/multifunctionality aspects of Irish agriculture. The analysis also provides projections of the potential effects of a WTO agreement on such measures in Ireland.

Under baseline policies (the Luxembourg CAP Reform Agreement), GHG emissions from Irish agriculture are projected to decline over the next 10 years relative to existing levels. Potential WTO trade reforms that might arise from a future WTO agreement would lead to only modest additional reductions in GHG emissions by 2015.

In Ireland, increasing milk yields in the presence of a milk quota and the introduction of decoupled payments will reduce the number of dairy cows, other cattle and sheep. These kinds of livestock are the three leading contributors to GHG emissions from Irish agriculture. As a result of EU CAP reform (the decoupling of direct payments) and ongoing productivity improvements in agriculture, substantial reductions in methane and nitrous oxide emissions are possible, even in the absence of trade reform.

The dairy sector will remain the main source of agricultural GHG emissions in Ireland. This sector is projected to continue to produce at the maximum level allowed under the milk quota system. It is likely that it would require a greater degree of WTO reform than that examined in this WTO scenario to significantly reduce emissions from the dairy sector below the level projected in the baseline.

Estimates for 2004 indicate that agriculture was responsible for over one-quarter of all Irish GHG emissions. Consequently, the reduction in emissions from Irish agriculture stemming from both the CAP reform of 2003 and any future WTO trade reform should represent a significant contribution from the agricultural sector towards meeting the Irish national Kyoto target for the first commitment period of a maximum 13% increase in GHG emissions over the 1990 emissions levels.

Emissions of ammonia from Irish agriculture are projected to decline under both the baseline and the WTO reform scenarios. Yet much of the decline projected under the WTO scenario is estimated to occur in any event under the baseline scenario. The impact of the WTO scenario on ammonia emissions from agriculture only represents an additional 1% reduction by 2015, relative to the 2015 baseline position.

It should be noted that the projections in this study have been generated at a national level only. While this national focus does present an issue in the case of GHG emissions, in the case of ammonia emissions there may be additional regional or local considerations that fall outside the scope of this model. For example, while it is anticipated that trade reform will lead to an overall lowering of agricultural output in Ireland, it could lead to local-level intensification or extensification of production, which the national model is unable to capture. Local-level changes of these kinds could have local-level environmental implications concerning ammonia emissions that cannot be measured by a non-local level study such as this.

The projections in this chapter have been produced under the IPCC Tier I basis, since this is the level of detail allowed by the FAPRI-Ireland commodity model as currently structured. Future work will aim at redesigning aspects of the FAPRI-Ireland commodity model to allow a greater disaggregation of agricultural activity and enable emissions projections to be made on an IPCC Tier II basis.

Bibliography

Behan, J. and B. Hyde (2003), *Baseline Projections of Ammonia Emissions from Irish Agriculture*, Rural Economy Research Centre, Teagasc, Ireland.

Binfield, J., T. Donnellan, K. Hanrahan and P. Westhoff (2006), "The World Trade Reform: Possible Impact of the Doha Round on EU and Irish Agriculture", mimeo, Teagasc, Ireland.

Department of the Environment, Heritage and Local Government (1997), *Ireland Second National Communication under the United Nations Framework Convention on Climate Change*, Dublin.

————— (1998), *Limitation and Reduction of CO$_2$ and Other Greenhouse Gas Emissions in Ireland*, Dublin.

————— (2000), *National Climate Change Strategy Ireland*, Stationery Office, Dublin.

European Commission (2002), Communication on the Mid-Term Review of the Common Agricultural Policy, COM(2002) 394 final, Brussels.

European Council (2001), Directive 2001/81/EC of 23 October on national emission ceilings for certain atmospheric pollutants, OJ L 309/22, 27.11.2001.

Environmental Protection Agency (EPA) (2005), *Ireland's Environment 2004 – The State of the Environment*, EPA, County Wexford, Ireland (http://www.epa.ie/NewsCentre/ReportsPublications/ IrelandsEnvironment2004).

Houghton, J.J., L.G. Meiro Filho, B.A. Callander, N. Harris, A. Kattenberg and K. Maskell (eds) (1996), *Climate Change 1995: The Science of Climate Change, Contribution of Working Group I to the Second Assessment Report of the Intergovernmental Panel on Climate Change*, Cambridge and New York: Cambridge University Press.

Intergovernmental Panel on Climate Change (IPCC) (1997), *Revised 1996 IPCC Guidelines for National Greenhouse Gas Inventories: Reference Manual*, IPCC, Geneva (http://www.ipcc-nggip.iges.or.jp/public/ gl/invs1.htm).

————— (2001), *IPCC Good Practice Guidance and Uncertainty Management in National Greenhouse Gas Inventories*, IPCC, Geneva (http://www. ipcc-nggip.iges.or.jp/public/gp/gpgaum.htm).

OECD (2001), *Multifunctionality: Towards an Analytical Framework*, OECD, Paris, ISBN 9264186255.

United Nations (1992), United Nations Framework Convention for Climate Change, United Nations, New York (http://unfccc.int/resource/ ccsites/senegal/conven.htm).

United Nations Economic Commission for Europe (UNECE) (1999), Gothenburg Protocol to the 1979 Convention on Long-Range Transboundary Air Pollution (CLRTAP) on the Abatement of Acidification, Eutrophication and Ground-level Ozone, UN, New York.

US Department of State, Bureau of Oceans and International Environmental and Scientific Affairs (1998), *The Kyoto Protocol on Climate Change,* Washington, D.C. (http://www.state.gov/www/ global/oes/fs_kyoto_climate_980115.html).

US Environmental Protection Agency (EPA), Global warming website, EPA, Washington, D.C. (http://www.epa.gov/global warming).

10. Modelling the effects of trade policy on the multifunctionality of agriculture
A social accounting matrix approach for Greece

Demetrios Psaltopoulos and Eudokia Balamou

Introduction

During the last two decades, there have been major changes in the economic structure of European rural areas, mainly induced by agricultural policy reforms, international trade liberalisation and globalisation, and the strengthened role of rural development policies. The part played by agriculture and farmers in the European economy and society has been changing accordingly. Agricultural activity has suffered a setback in economic and social terms and especially with regard to employment, while manufacturing and service employment have spread.

At the same time, and especially in Europe, a number of concerns such as food over-production (which has led to major trade disputes and the negative and positive environmental consequences of modern farming) have prompted a rethinking of the position of agriculture within society. This trend has been accompanied by a reconsideration of the institutional system surrounding agriculture (van Huylenbroeck & Durand, 2004) and consequently, a greater emphasis on the non-production functions of farming. Against this background and (particularly) in view of the Doha round of World Trade Organisation (WTO) talks, increased attention has been given to the concept of 'multifunctionality', which is based on the notion that agriculture produces multiple outputs that include both public goods and privately traded commodities (OECD, 1997; Peterson, Boisvert & de Gorter, 2002). Furthermore, in the EU, this concept has been utilised by several European Commission statements defending the right of the EU to use common agricultural policy (CAP) support in order to uphold a multifunctional 'European model of agriculture' (European Commission, 1998; Thomson, 2004).

Nowadays, over half of the population in the EU-25 live in rural areas, which cover 90% of the EU's territory. Despite the fact that rural areas have increasingly become an environment for living and leisure, farming and forestry are still of overriding importance for land use and the management of natural resources, and consequently, a major platform for economic development in rural communities. Thus any changes in agricultural support induced by international trade negotiations would surely influence farm activities and the joint production of both public and private goods by agriculture. Taking into account that most of these commodity and non-commodity outputs are directly and indirectly 'traded' (Bryden, 2005), it is interesting to explore the multifunctionality implications of alternative trade-policy scenarios.

The objective of this chapter is to utilise the social accounting matrix (SAM) analytical framework for conducting an assessment of the potential effects of trade agreements on several multifunctionality indicators in Greek agriculture. More specifically, two SAM models are constructed, one for Greece and one for the local rural economy of Archanes (Crete), an agriculturally dependent NUTS IV area that has demonstrated a noticeable record in terms of the implementation of pillar II policies. The alternative scenarios considered here were specified by Dwyer et al. (2005). Based on this specification and pertinent decisions about CAP reform and pillar II policies (Council of the EU, 2003; European Commission, 2004a and 2004b), a national/regional specification of these scenarios is followed by SAM-based impact analyses. Scenario-specific effects include estimates of annual average changes in agricultural output, employment and land use, economy-wide output and employment, factor incomes (labour and capital), household and firm income and finally, pollution emissions.

This chapter is organised as follows. Section 10.1 presents the applied methodology and the model construction process. Section 10.2 presents the methodological procedures of the policy impact assessment, while section 10.3 presents the specification of the alternative policy scenarios and the estimation of effects. The chapter ends with relevant conclusions.

10.1 Modelling framework

10.1.1 Methods

The repercussions of rural and agricultural policies have been evaluated by different tools and approaches, as regards targeted groups in various rural areas (Bossard, Daucé & Léon, 2000). Quantitative evaluations range from descriptive techniques, rational checking procedures and local growth indicators, to more sophisticated macro- and micro- models, input–output (IO) models, cost-benefit and multi-criteria analyses (for a review, see Psaltopoulos, 2004). Several studies, however, have used some form of qualitative analysis to evaluate rural policy action. Evaluation of CAP effects has also taken a number of directions, such as emphasising environmental or competitive aspects, and has become part of the overall regional analysis in the Cohesion Reports of the European Commission (1996, 2001 and 2004c).

The selection of an 'appropriate' evaluation technique mainly depends on the policy actions to be evaluated and on the focus of the evaluation. As policy interventions are made at distinct levels and as policy is usually defined as "a set of activities which may differ and may have different direct beneficiaries at different domains, and which are directed towards common general objectives or goals" (European Commission, 1997), a general equilibrium approach seems more appropriate for assessing policy impacts. Such models, based on the SAM technique, allow the identification of the effects of both pillar I and II funding (i.e. investment and direct income transfers) in a national or local economy (or both). Other possible advantages of this modelling framework can be described as follows:

a) The multi-sectoral dimension of a SAM approach accommodates the analysis of the effects induced by current rural development policies, which have shifted attention from traditional, product-/sector-oriented support towards more broadly based (multi-sectoral) assistance.

b) Several evaluation approaches only estimate the direct effects of policy action. As a general equilibrium approach, however, the SAM technique allows for the estimation of the 'global' economic effects of these injections.

c)   The SAM technique (in comparison with the more traditional Leontief IO approach) has the ability to capture the distributional effects of exogenous injections (investment funding and transfers) in an economy. More specifically, the important presence of the CAP subsidy payments to farmers further substantiates the use of the SAM method since analysis that focuses solely on production linkages (IO) may ignore the implications (particularly the distributional effects) arising from other types of links between rural sectors (especially agriculture) and the macroeconomy.

On the other hand, it has to be noted that the use of this technique for impact assessment also involves some simplistic assumptions regarding the economic behaviour of sectors, households, etc., which are all assumed to maintain their recorded pattern of expenditure in the base period (the linearity assumption). Also, the 'snapshot' nature of the technique does not allow the exploration of changes in technology, relative prices, incomes and expenditures over time. Most of the above weaknesses could have been dealt with here through the use of computable, general equilibrium analysis (and a considerable number of additional and often speculative assumptions), but this is clearly beyond the resources of this effort.

The SAM technique has not often been used for policy-impact analysis, mainly because of (usually) severe data demands, especially at the regional level. But in recent years some indicative studies have applied this technique. Psaltopoulos (2001) estimated the economy-wide effects of alternative policy scenarios related to the tobacco sector on the national economy of Greece. Roberts (2003) built a 1997 SAM for the Western Isles in Scotland and estimated the economic impact of both central government funding of public services and exogenous transfers of income to local households. Psaltopoulos et al. (2004) built regional SAMs for six remote rural areas of Scotland, Finland and Greece, in an attempt to discover how EU structural policies have affected their economies. Finally, Psaltopoulos & Balamou (2005) built an interregional SAM for Crete to assess the effects of CAP pillar I and II measures on a rural–urban interregional economy.

10.1.2 Application

The objective of this section is to present the analytical procedure applied to the generation of a national SAM for the Greek economy and a regional SAM for the local rural economy of Archanes, in both cases for the year 1998.

The national SAM for Greece

The basis of the national SAM is the national accounts and a detailed IO table for 1998. Building on this foundation and by using the statistical tables of the National Statistical Service of Greece (NSSG) we have been able to add the accounts of various economic agents. These accounts were drawn from a national household income and expenditure survey, the financial accounts of enterprises, the rest of the world and non-profit institutions, the principal aggregates of the Eurostat national accounts, transactions table, data on taxes, subsidies, government transfers, etc.

The initial product of the above process was a preliminary 17-sector SAM for Greece for 1998. The matrix contained aggregated structural information for agriculture and forestry. Therefore, in order to generate detailed information on the structure of sectors relevant to this study we disaggregated agriculture and forestry into the cultivation of arable crops, vegetables, fruit, tobacco, livestock and forestry. As a result, the final form of the national SAM consists of twenty-two sectors, two production factors (labour and capital), three institutions (households, firms and government), the rest of the world and a capital account.

The regional SAM for Archanes

The regional IO table for Archanes has been generated by using the hybrid GRIT technique developed by Jensen et al. (1979). The GRIT technique creates a preliminary, regional transactions matrix through the mechanical adjustment of the national direct-requirements matrix by using employment-based simple location quotients and cross-industry location quotients. Subsequently, an analyst can 'interfere' with the mechanically-produced table through the insertion of 'superior' data from surveys or other sources at various stages in the development of the table. Thus, GRIT incorporates the advantages of both the 'survey' and 'non-survey' IO regionalisation approaches.

After regionalising the available national IO table (first, to the level of the prefecture and then to that of the study area) with the use of the mechanical GRIT procedure, information from a sectoral business survey in Archanes was utilised. The selection of target sectors for the business survey was primarily based on the importance of particular sectors in the structure of the local economy and in the implementation of pillar I and II policies. Businesses were primarily chosen through random sampling of business directories supplied by local authorities. In some cases, major firms were deliberately selected owing to their significant economic impact on the study area (which mostly consists of small enterprises). Surveys were conducted face-to-face with business owners, using a structured questionnaire. The sample accounted for 40% of local units. The second main source of superior data was an extended survey of households in Archanes. Around 10% of local households provided information on the sources of their income and their consumption patterns. In order to develop the non-IO components of the regional SAM, regional and national data from a wide range of sources were used (the 1998 Household Income and Expenditure Survey, the NSSG regional accounts, interviews with local policy-makers and local government data).

As a result, the final form of the Archanes SAM consists of thirteen sectors (with three agricultural sub-sectors – vine-growing, olive-growing and other agriculture), three production factors (labour, capital and land), three institutions (households, firms and government), the rest of the world and a capital account.

With regard to the identification of multifunctionality indicators, these models estimate changes in farm output levels, agricultural employment, total employment, agricultural land use and pollution emissions. Finally, estimated results are presented in an average annual form for the period 2007–13, with the exception of tobacco, where estimates relate to the post–2010 period.

10.2   Impact analysis methodology

10.2.1 Conceptual issues

In accordance with the SAM analytical framework (general equilibrium, comparative static), impact analysis deals with the comparison of levels of study-area output, employment and so forth, calculated by applying multiplier and coefficient values to the injections of expenditure (treated as additional final demand) associated with policy. Implicitly, this compares two alternative equilibrium positions of the national/regional economy (i.e. mutually balanced levels of production, firm and household incomes, and trade flows), which are consistent respectively with and without these expenditure patterns. No account is taken of the time pattern of adjustment to the additions to final demand, while calculations seek to isolate the effects of policy expenditures from those of other influences.

In this particular exercise scenario-specific domestic policy changes have been fed into the SAM models as injections to final demand. More specifically,

- A decrease in subsidies (owing to modulation and other revisions to CAP market support) constitutes a negative injection to the agricultural subsidies cell (government column).

- Any increase in pillar II funds is converted into projections of rural development action (programmes and measures) and constitutes an increase in the relevant capital account column (for investment action) or agricultural subsidies cell (for agri-environment measures). In terms of the sectoral distribution of these changes to exogenous final demand, the 2000–06 pattern is observed in both study areas.

- Possible adjustments to production volumes and their effects (e.g. an increase in the milk quota) can be modelled through the use of the mixed endogenous/exogenous version of the Leontief model, extended to a SAM framework.

- In the case of a decline in prices, supply is linearly adjusted through the use of the relevant product-specific price elasticities (obtained from Mergos, 2003) and the effects are modelled through the use of the mixed endogenous/exogenous version of the Leontief model, extended to a SAM framework.

- The effects of possible substitution (e.g. moving from cotton to cereals) or abandonment of agricultural activity in several sub-sectors (or both) owing to decoupling are modelled through the use of the mixed endogenous/exogenous version of the Leontief model, extended to a SAM framework. Exogenous estimates of these developments are obtained from Tsiboukas (2003).

- Finally, the impact on land use is projected through the utilisation of the applicable input elasticities estimated in Sarris (2003).

10.2.2 Analytical procedures

Conventional Leontief procedure

In a SAM framework, the conventional Leontief procedure can be used to estimate the economy-wide effects of changes in exogenous demand. More analytically, the identification of the shocks whose effects are investigated (e.g. changes in investment from pillar II measures, in consumer demand or in pillar I subsidies) is followed by the specification of the model's exogenous accounts (in this case the government, the rest of the world and capital). The available SAM multipliers and coefficients are then utilised in order to produce economy-wide effects in terms of output, labour income, firm income, household income and employment.

Economic effects of fixed supply

In parallel, in a SAM context, exogenous changes to sectoral gross output(s) as a result of forces outside the model can have a profound impact on the accounts of the other aspects of the economy under study, through the interdependence relationships portrayed by this general equilibrium data system. Such changes include a decline in prices (which causes an adjustment in supply), the abandonment or shift in cultivation as a result of decoupling, an increase of production quotas and targets, and natural disasters.

This method is based on the mixed exogenous/endogenous variable version of the IO model devised by Miller & Blair (1985) for IO analysis. It was extended to a SAM context by Roberts (1992), who estimated the (UK) economy-wide effects of milk quotas, which represent a ceiling on the level of gross output of a particular sector. It was further extended by Psaltopoulos & Thomson (1998) to estimate the capacity-adjustment effects of structural policy implementation in remote rural areas of the EU. Through these methods, changes in activity levels lead to the execution of comparative analysis and the estimation of the relevant economic effects (of changes in supply) on output, labour income, capital income, land rent (for the regional model), firm income, household income and employment.

Impact on pollution emissions

In order to estimate the scenario-specific levels of pollution emissions, we have drawn from the methodology developed by Leontief & Ford (1972) and utilised the national pollution matrix produced by the NSSG (Mylonas, 2000).

This particular matrix was transformed by Loizou (2001) to reflect the disaggregation of agriculture into several sub-sectors. As a next step, a matrix of total pollution coefficients was estimated for both the national (Greece) and regional (Archanes) economies, after carrying out the appropriate sectoral classification adjustments. Elements in this matrix indicate the total (i.e. direct and indirect) pollution of pollutant k, which occurs from increased economic activity in sector j, caused (in turn) by a unitary increase of final demand within this particular sector. The emissions estimated concerned nine pollutants (CO_2, CH_4, N_2O, NOx, CO, NMVOC, SO_2, BOD_5 and nitrates).

10.3   Scenario analysis

During the period in which this study was undertaken, an important issue concerned the determination by the Greek Ministry of Agriculture of the degree of decoupling in various regimes such as arable crops and sheep farming, along with the distribution of the national envelope in the case of (e.g.) olive oil. In order to proceed with the specification of the scenarios, we utilised information provided by the Ministry of Agriculture (2004). This information was derived from a committee of experts established by the ministry in order to provide an opinion on the implementation of new CAP reforms in Greece. According to the committee, the main characteristics of the implementation of the CAP reforms in Greece would be as follows:

- A single payment scheme would be implemented for all products covered by the reform (which in fact commenced on 1.1.2006).

- Full decoupling would occur in the sectors covering arable crops (including durum wheat), sheep and goat farming, and olive oil.

- In the bovine sector, Greece would opt for keeping 100% of the suckler cow premium and 40% of the slaughter premium coupled.

- Greece would not utilise the regional application options.

- Greece would not utilise the option to grant up to 10% of the national ceiling as sector-specific payments for improving the quality and marketing of agricultural products.

- The single payment scheme would fully apply to the Aegean Islands.

In most instances these forecasts turned out to be accurate.

10.3.1 Specification of scenarios

Based on the decisions concerning the 2003–04 reforms to the CAP and pillar II policies (Council of the EU, 2003; European Commission, 2004a and 2004b), we carried out the national and regional specification of the alternative scenario elements.

For Greece, scenario 1bis consists of five scenario elements. The first element is subsidies.[1] This element will feature a decline in subsidies for durum wheat (by €12.4 mn per annum) and for cotton (by €12 mn), along with a fall of the reference period levels for rice (by 6.8%) and for tobacco (by 50%, post–2010). It will also involve a rise in subsidies for other cereals (by €9 mn) and for dairy products (by €29 mn).

The second scenario element is prices. In the case of rice, a 15% reduction in prices is projected, owing to assumptions about market liberalisation. Multiplied by the relevant supply elasticity, this results in a decrease of 18.6% in the gross production of rice. In dairy products a 10% decrease in gross output is projected.

In the third scenario element, modulation, there is a 2.4% decrease in subsidies for the products covered by the CAP reform. The fourth scenario element involves decoupling, which projects that 30% of cotton production is abandoned and converted to cereals. In this scenario, 40% of tobacco production is also abandoned, of which 11% is converted to sugar beet. Declines in gross output as a result of decoupling are anticipated for cereals (-14%), olive oil (-10%), sheep and goat farming (-15%) and beef farming (-5%). Finally, in the case of pillar II policies, the annual average expenditure for the period 2000–06 (which includes all European Agricultural Guidance and Guarantee funding for this purpose) is projected to increase by €88.6 mn of Community contributions, related to modulation and sectoral transfers.

For Archanes, scenario 1 consists of three scenario elements. The first one is modulation. For this element, when account is taken that 30% of olive oil farms (and 55% of specific subsidies for this product) relate to farmers over the €5,000 threshold, a cut in subsidies (2.8%) is expected for this particular product. For the second scenario element, decoupling, 20% of olive oil production is assumed to be abandoned. In the third scenario element, which considers pillar II policies, an increase of average 2000–06 annual expenditures by 25% is projected.

Scenario 2 consists of the above three elements plus the element of prices. More specifically, in olive oil a 10% decrease in prices is projected as a result of market liberalisation. Taking into account the relevant supply

---

[1] These elements are drawn from Bourdaras (2004 and 2005).

elasticity, this decline leads to a 2.56% reduction in gross output. Similarly, declines in gross output are anticipated for raisins (by 1.46%), grapes for wine (by 4.2%) and table grapes (by 3.17%).

Scenario 2b consists of all the scenario 2 elements plus the element of subsidies, wherein olive oil is assumed to suffer a cut of 20% in subsidies (in reference period levels).

Scenario 3 consists of all the scenario 2b elements, plus the element of pillar II policies. This element expects an increase in policy expenditures that is equivalent to the decline in olive oil subsidies.

10.3.2 *Results of the scenario effects*

Tables 10.1-10.2 and 10.3-10.6 present the estimated effects of the above scenarios for Greece and Archanes, respectively. The tables include results for several variables, some of which may not be linked to multifunctionality. Comments in this section, however, do highlight projected changes to indicators that are related to the multifunctionality concept.

In the case of Greece, it seems that if scenario 1bis is realised, the impact on the national economy will be marginally positive, with the exception of agricultural employment, for which a decline of 10.11% is forecasted (Table 10.1).

Agricultural output is expected to decline mainly in the case of tobacco (-38.99%), as well as in livestock (-5.49%) and fruit (-3.22%). Output in the vegetables sector is forecasted to increase by 1.15%, while the output of arable crops remains more or less constant. Farm employment is projected to fall by a significant 10.11%, mainly owing to developments in the tobacco sector. In terms of land use, a 10.3% reduction of tobacco land is projected, while the livestock figure is expected to decline by 1.73%. At the economy-wide level, output effects seem to be positive (+0.66%) through the increase of pillar II spending and the declining importance of agriculture in the Greek economy, while (for the same reasons) a moderate increase in total employment is projected (+0.10%). Finally, pollution emissions are expected to increase by 1.97%, a figure attributed to a projected rise of 5.45% as a result of pillar II policies and a decline of 3.48% through the contraction of farm activity.

Table 10.1 Effects of scenario 1bis for Greece (annual average changes from 1998 levels, 1998 prices)

Scenario elements	Output effects (mn of GRD)	Change (%)	Labour income effects (mn of GRD)	Change (%)	Capital income effects (mn of GRD)	Change (%)	Firm income effects (mn of GRD)	Change (%)
A. Decline in subsidies	-46,972.54	-0.07	-7,329.99	-0.06	-64,367.51	-0.32	-61,571.41	-0.32
B. Decline in prices	-65,549.41	-0.10	-5,914.43	-0.05	-33,200.80	-0.16	-31,758.57	-0.16
C. Modulation	-15,730.18	-0.02	-2,454.67	-0.02	-21,555.42	-0.11	-20,619.06	-0.11
D. Decoupling	-325,627.10	-0.52	-46,553.68	-0.39	-104,178.69	-0.52	-99,653.20	-0.51
E. Pillar II	868,512.04	1.38	13,340.20	1.16	397,586.45	1.97	380,315.39	1.96
Total	414,632.81	0.66	77,087.42	0.64	174,284.02	0.87	166,713.17	0.86

Scenario elements	Household income effects (mn of GRD)	Change (%)	Agricultural emp. effects (no. of jobs)	Change (%)	Employment effects (no. of jobs)	Change (%)	Pollution effects (tonnes)	Change (%)
A. Decline in subsidies	-46,020.65	-0.14	-551	-0.08	-10,624	-0.17	-248,708.99	-0.30
B. Decline in prices	-25,888.48	-0.08	-4,229	-0.61	-20,401	-0.34	-409,075.75	-0.49
C. Modulation	-15,411.42	-0.05	-184	-0.03	-3,588	-0.06	-83,287.77	-0.10
D. Decoupling	-109,457.08	-0.34	-74,640	-10.84	-144,759	-2.38	-2,179,302.14	-2.60
E. Pillar II	379,071.78	1.17	9,988	1.45	185,588	3.05	4,570,732.83	5.45
Total	182,314.15	0.56	-69,616	-10.11	6,246	0.10	1,650,335.18	1.97

Source: Authors' calculations.

In terms of the elements of scenario 1bis, it should be noted that falls in agricultural output (-5.3%), employment (-10.84%) and total employment (-2.38%) mostly occur from decoupling, which also contributes to a 2.6% reduction of pollution. The effects of the remaining elements of the scenario (declines in subsidies and prices, and modulation) seem to be rather marginal, while developments in pillar II policies generate positive outcomes even in terms of farm output (+1.64%) and farm employment (+1.45%), but more importantly, for total employment (+3.05%).

The forecasts show that pollution emissions may increase in total, but there are projections (in Table 10.2) of reductions in the cases of BOD_5 (-7.56%), N_2O (-7.54%), CH_4 (-4%) and nitrates (-2.30%). On the other hand, these positive projections are rather 'eliminated' by an estimated 2.08% increase in CO_2 emissions, the particular pollutant that is by far the most important in Greece.

Table 10.2 *Effects of scenario 1bis for Greece on pollution (annual average changes from 1998 levels)*

Pollutants	Changes in pollution (tonnes)	Change (%)
CO_2	1,680,069.28	2.08
CH_4	-13,517.87	-4.00
N_2O	-2,125.88	-7.54
Nox	2,991.64	0.99
CO	2,645.42	0.30
NMVOC	763.74	0.29
SO_2	11,502.17	2.19
BOD_5	-22,925.86	-7.56
Nitrates	-9,047.48	-2.30
Total	1,650,355.18	1.97

Source: Authors' calculations.

For the agriculturally-dependent local economy of Archanes, scenario projections are rather negative. Not surprisingly, scenario 2b (a reduction of income support aids) generates the most negative results, followed by scenario 2 (full decoupling and the elimination of export subsidies) and scenario 3. The fact that the status quo-specific scenario 1 seems to be associated with (comparatively) less-pessimistic prospects is possibly a welcomed consolation.

As in the case of national projections, the estimated negative effects are quite significant for farm output and employment, being around the -11% mark for scenarios 2, 2b and 3. Furthermore, economy-wide job losses are notable for these three scenarios (ranging from -3.8% to -6.6%), as are reductions in agricultural land use (around -8%). Projections on the reduction of pollution are also quite significant for these three scenarios, ranging from -2.14% for scenario 3 to more than -5% for scenarios 2 and 2b.

In more detail, in the case of the status quo scenario 1 (Table 10.3), agricultural output is expected to fall by 5.9%, mainly owing to the decline in olive oil production (-18.6%). For the same reason, farm employment is projected to fall by 5.22%. In terms of land use, a 5.44% reduction of land dedicated to olive trees is anticipated, while land dedicated to vineyards could increase by 0.5%. At the local economy-wide level, output effects seem to be marginally positive (+0.03%) as a result of the increase in pillar II spending, while (for the same reasons) a rather moderate decline in total employment is projected (-1.88%). A cut of 0.35% in pollution emissions is expected, based on an estimated 3.55% increase arising from pillar II policies and a fall of 3.90% through the contraction of farm activity.

Examining the elements of scenario 1, it should be noted that declines in agricultural output (-6.6%) employment (-6.55%) and total employment (-4.74%) are related to decoupling, which also contributes to a 3.86% fall in pollution. The impact of modulation seems to be marginal. Changes in pillar II policies, however, generate positive results even in the cases of farm employment (+1.33%) and total employment (+2.91%). Finally, pollution forecasts show a decline in total emissions, but this projection is almost solely attributable to a projected reduction of nitrates (-6.05%).

The results associated with scenario 2 (the elimination of export subsidies) are even more negative, owing to the repercussions of the projected decline in prices. Agricultural output is expected to decline (Table 10.4) by a significant 11.3%, as olive oil production is reduced by 21.17% and vine production by 8.33%. For the same reasons, farm employment is projected to fall by a significant 11.4%. With regard to land use, a 6.18% reduction of land dedicated to olive trees is anticipated, while land dedicated to vineyards could decline by 1.86%.

Table 10.3 Effects of scenario 1 for Archanes (annual average changes from 1998 levels, 1998 prices)

Scenario elements	Output effects (mn of GRD)	Change (%)	Labour income effects (mn of GRD)	Change (%)	Capital income effects (mn of GRD)	Change (%)	Land rent effects (mn of GRD)	Change (%)	Firm income effects (mn of GRD)	Change (%)
A. Modulation	-5.50	-0.03	-0.74	-0.02	-11.82	-0.17	-0.50	-0.03	-11.88	-0.14
B. Decoupling	-450.62	-1.91	-36.19	-0.86	-110.47	-1.62	-106.89	-5.60	-209.50	-2.43
C. Pillar II	463.80	1.96	65.03	1.55	587.06	8.61	31.68	1.66	596.35	6.91
Total	7.68	0.03	28.10	0.67	464.74	6.81	-75.71	-3.96	374.97	4.34

Scenario elements	Household income effects (mn of GRD)	Change (%)	Agricultural emp. effects (no. of jobs)	Change (%)	Employment effects (no. of jobs)	Change (%)	Pollution effects (tonnes)	Change (%)
A. Modulation	-9.03	-0.07	0	0	-1	-0.05	-3.32	-0.04
B. Decoupling	-181.00	-1.32	-56	-6.55	-93	-4.74	-335.98	-3.86
C. Pillar II	479.51	3.49	11	1.33	57	2.91	308.79	3.55
Total	289.49	2.11	-45	-5.22	-37	-1.88	-30.51	-0.35

Source: Authors' calculations.

Table 10.4 Effects of scenario 2 for Archanes (annual average changes from 1998 levels, 1998 prices)

Scenario elements	Output effects (mn of GRD)	Change (%)	Labour income effects (mn of GRD)	Change (%)	Capital income effects (mn of GRD)	Change (%)	Land rent effects (mn of GRD)	Change (%)	Firm income effects (mn of GRD)	Change (%)
A. Modulation	-5.50	-0.03	-0.74	-0.02	-11.82	-0.17	-0.50	-0.03	-11.88	-0.14
B. Decoupling	-450.62	-1.91	-36.19	-0.86	-110.47	-1.62	-106.89	-5.60	-209.50	-2.43
C. Pillar II	463.80	1.96	65.03	1.55	587.06	8.61	31.68	1.66	596.35	6.91
D. Decline in prices	-419.59	-1.77	-32.74	-0.78	-107.99	-1.58	-93.27	-4.88	-193.99	-2.25
Total	-411.91	-1.74	-4.64	-0.11	356.78	5.24	-168.98	-8.85	180.98	2.09

Scenario elements	Household income effects (mn of GRD)	Change (%)	Agricultural emp. effects (no. of jobs)	Change (%)	Employment effects (no. of jobs)	Change (%)	Pollution effects (tonnes)	Change (%)
A. Modulation	-9.03	-0.07	0	0	-1	-0.05	-3.32	-0.04
B. Decoupling	-181.00	-1.32	-56	-6.55	-93	-4.74	-335.98	-3.86
C. Pillar II	479.51	3.49	11	1.33	57	2.91	308.79	3.55
D. Decline in prices	-166.88	-1.22	-53	-6.18	-86	-4.39	-429.51	-4.93
Total	122.60	0.88	-98	-11.40	-123	-6.27	-460.01	-5.29

Source: Authors' calculations.

Table 10.5 Effects of scenario 2b for Archanes (annual average changes from 1998 levels, 1998 prices)

Scenario elements	Output effects (mn of GRD)	Change (%)	Labour income effects (mn of GRD)	Change (%)	Capital income effects (mn of GRD)	Change (%)	Land rent effects (mn of GRD)	Change (%)	Firm income effects (mn of GRD)	Change (%)
A. Modulation	-5.50	-0.03	-0.74	-0.02	-11.82	-0.17	-0.50	-0.03	-11.88	-0.14
B. Decoupling	-450.62	-1.91	-36.19	-0.86	-110.47	-1.62	-106.89	-5.60	-209.50	-2.43
C. Pillar II	463.80	1.96	65.03	1.55	587.06	8.61	31.68	1.66	596.35	6.91
D. Decline in prices	-419.59	-1.77	-32.74	-0.78	-107.99	-1.58	-93.27	-4.88	-193.99	-2.25
E. Decline in subsidies	-39.26	-0.17	-5.27	-0.13	-84.46	-1.24	-3.57	-0.19	-84.84	-0.98
Total	-451.17	-1.91	-9.91	-0.24	272.32	4.00	-172.55	-9.04	96.14	1.11

Scenario elements	Household income effects (mn of GRD)	Change (%)	Agricultural emp. effects (no. of jobs)	Change (%)	Employment effects (no. of jobs)	Change (%)	Pollution effects (tonnes)	Change (%)
A. Modulation	-9.03	-0.07	0	0	-1	-0.05	-3.32	-0.04
B. Decoupling	-181.00	-1.32	-56	-6.55	-93	-4.74	-335.98	-3.86
C. Pillar II	479.51	3.49	11	1.33	57	2.91	308.79	3.55
D. Decline in prices	-166.88	-1.22	-53	-6.18	-86	-4.39	-429.51	-4.93
E. Decline in subsidies	-64.47	-0.47	-2	-0.18	-6	-0.31	-23.69	-0.27
Total	58.13	0.41	-100	-11.58	-129	-6.58	-483.71	-5.55

Source: Authors' calculations.

Table 10.6 Effects of scenario 3 for Archanes (annual average changes from 1998 levels, 1998 prices)

Scenario elements	Output effects (mn of GRD)	Change (%)	Labour income effects (mn of GRD)	Change (%)	Capital income effects (mn of GRD)	Change (%)	Land rent effects (mn of GRD)	Change (%)	Firm income effects (mn of GRD)	Change (%)
A. Modulation	-5.50	-0.03	-0.74	-0.02	-11.82	-0.17	-0.50	-0.03	-11.88	-0.14
B. Decoupling	-450.62	-1.91	-36.19	-0.86	-110.47	-1.62	-106.89	-5.60	-209.50	-2.43
C. Pillar II	463.80	1.96	65.03	1.55	587.06	8.61	31.68	1.66	596.35	6.91
D. Decline in prices	-419.59	-1.77	-32.74	-0.78	-107.99	-1.58	-93.27	-4.88	-193.99	-2.25
E. Decline in subsidies	-39.26	-0.17	-5.27	-0.13	-84.46	-1.24	-3.57	-0.19	-84.84	-0.98
F. Increase in pillar II funds	445.25	1.88	62.43	1.49	563.58	8.26	30.41	1.59	572.49	6.63
Total	-5.91	-0.03	52.53	1.25	835.89	12.25	-142.13	-7.44	668.63	7.74

Scenario elements	Household income effects (mn of GRD)	Change (%)	Agricultural emp. effects (no. of jobs)	Change (%)	Employment effects (no. of jobs)	Change (%)	Pollution effects (tonnes)	Change (%)
A Modulation	-9.03	-0.07	0	0	-1	-0.05	-3.32	-0.04
B. Decoupling	-181.00	-1.32	-56	-6.55	-93	-4.74	-335.98	-3.86
C. Pillar II	479.51	3.49	11	1.33	57	2.91	308.79	3.55
D. Decline in prices	-166.88	-1.22	-53	-6.18	-86	-4.39	-429.51	-4.93
E. Decline in subsidies	-64.47	-0.47	-2	-0.18	-6	-0.31	-23.69	-0.27
F. Increase in pillar II funds	460.33	3.35	11	1.27	54	2.75	296.44	3.41
Total	518.47	3.78	-89	-10.31	-75	-3.83	-187.27	-2.14

Source: Authors' calculations.

At the economy-wide level, output effects seem to be negative (-1.74%), because of the negative effects of both modulation and the expected decline in prices, while (for the same reasons) projections are also negative for total employment (-6.27%). Pollution emissions are expected to fall by an important 5.29%, as the reduction in prices under this scenario could contribute to a decline of 4.93%.

In scenario 2 the contributions of decoupling and the expected decline in prices to the forecasted negative trends seem to be rather balanced. Yet it appears that decoupling negatively affects the olive oil sub-sector the most, while a possible decline in prices seems to mostly hit vine-growers. The effects of modulation and pillar II policies are similar to those of scenario 1. Estimates of pollution emissions show a decline in almost all categories of emissions and especially in nitrates (-14.4%), CH_4 (-12.91%), N_2O (-12.46%) and even CO_2 (-4.51%).

The results for scenario 2b (further reductions in support) are more pessimistic than for scenario 2, owing to the marginally negative effects of a further decline in support (Table 10.5). Nevertheless, in terms of 'structural characteristics', the relevant projections are somewhat similar to those in scenario 2, as the negative contribution of a further cut in subsidies is rather marginal in all categories of estimates.

Finally, the increase in pillar II funds associated with scenario 3 improves the projections, especially for economy-wide output and employment (Table 10.6). On the other hand, estimates for agricultural output, agricultural employment and land abandonment differ only marginally from those of scenarios 2 and 2b. The anticipated decline of pollution emissions is cut by half (compared with the levels of scenarios 2 and 2b), as pillar II action seems to be associated with an increase in emissions (of all types).

Conclusions

Overall, the results of this analysis suggest that under the scenarios examined, the effects of policy reform upon multifunctionality indicators are rather mixed and perhaps not extremely worrying. The effects of the status quo scenarios seem to be optimistic in terms of projected economy-wide output and employment at both the national and regional levels. On the other hand, scenario 1(bis) generates negative results for farm output and employment (especially for the agriculturally-dependent Archanes economy), while projections of land-use abandonment are marginal at the national and rather moderate at the regional level, and environmental repercussions are negative at the national level. The regional analysis has also shown that (at least in this case study) the results from scenarios 2, 2b and (even) 3 are rather worrying in terms of all categories of projections, with the notable exception of the important one of economy-wide output. Taking into account the specification of scenario 3, this finding generates rather justified reservations about the ability of pillar II policies to ameliorate the contraction of economic activity caused by a decrease in pillar I support in such an agriculturally-dependent local economy.

References

Bossard, P., P. Daucé, and Y. Léon (2000), *The Effects of Rural Development Policy on Local and Regional Economic Growth*, paper presented at the "International Conference on European Rural Policy at the Crossroads", held in Aberdeen on 29 June–1 July 2000.

Bourdaras, D. (2004), *Mid-Term Review of the CAP: Summary and Comments of the Commission Proposals,* Working Document.

———— (2005), *The Support of Greek Agriculture during the Period 1989-1997,* Directorate for Agricultural Policy, Ministry of Agriculture, Hellenic Republic, Athens.

Bryden, J. (2005), *Multifunctionality, Rural Development and Policy Adjustments in the European Union,* paper presented at the seminar held by the Ministry of Agriculture, Fishery and Forestry Research in Tokyo on 16 February.

Council of the EU (2003), Interinstitutional File 2003/0006 (CNS), 26 September.

Dwyer, J., D. Baldock, H. Guyomard, J. Wilkin, and D. Klepacka (2005), *Scenarios for Modelling Trade Policy Effects on the Multifunctionality of European Agriculture,* ENARPRI Working Paper No. 10, CEPS, Brussels.

European Commission (1996), *First Report on Economic and Social Cohesion,* Office for Official Publications, European Commission, Brussels.

———— (1997), *Evaluation of EU Expenditure Programmes: A Guide,* Directorate-General for Budgets (DG XIX/02), European Commission, Brussels.

———— (1998), *Agenda 2000: Commission Proposals*, COM(1998) final, European Commission, Brussels.

————— (2001), *Second Report on Economic and Social Cohesion*, COM(2001) 24, European Commission, Brussels.

————— (2004a), *CAP Reform Continued: EU Agrees on More Competitive and Trade-Friendly Tobacco, Olive, Cotton and Hops Regimes*, European Commission, Brussels.

————— (2004b), *Proposal for a Council Regulation on Support for Rural Development by the European Agricultural Fund for Rural Development*, COM(2004) 490 final, European Commission, Brussels.

————— (2004c), *A New Partnership for Cohesion: Convergence, Competitiveness, Cooperation*, Third Report on Economic and Social Cohesion, European Commission, Brussels.

Jensen, R.C., T.D. Mandeville and N.D. Karunaratne (1979), *Regional Economic Planning*, London: Croom Helm.

Leontief, W. and D. Ford (1972), "Air Pollution and the Economic Structure: Empirical Results of Input-Output Computations", in A. Brody and A.P. Carter (eds), *Input-Output Techniques*, Amsterdam: North Holland, pp. 9-30.

Loizou, E.A. (2001), "A Quantitative Evaluation of the Impacts of the Production Process on the Environment", unpublished PhD thesis, Department of Agriculture, Aristotle University.

Mergos, G. (2003), *Supply Response in Greek Agriculture*, Department of Economics, University of Athens.

Miller, R.E. and P.D. Blair (1985), *Input-Output Analysis: Foundations and Extension*, Englewood Cliffs: Prentice Hall.

Ministry of Agriculture (2004), *Outcome of the Committee of Experts on the Implementation of the New CAP*, Ministry of Agriculture, Hellenic Republic, Athens.

Mylonas, N. (2000), "Presentation of the Greek NAMEA Tables for the Period 1988-1996", OECD Meeting of National Accounts Experts, 26-29 September, Paris.

OECD (1997), *Environmental Benefits from Agriculture: Issues and Policies*, OECD, Paris.

Peterson, J.M., R.N. Boisvert, and H. de Gorter (2002), "Environmental Policies for a Multifunctional Agricultural Sector in Open Economies", *European Review of Agricultural Economics*, Vol. 29, pp. 423-43.

Psaltopoulos, D. (2001), *The impacts of policy on the tobacco sector*, in D. Skuras (ed.), *Tobacco in the Greek Economy*, Athens: Gutenberg, pp. 213-34.

————— (2004), *Scenarios for Modelling Trade Impacts upon Multifunctionality in European Agriculture: The Greek Context*, ENARPRI Workpackage 3.3, ENARPRI, CEPS, Brussels.

Psaltopoulos, D. and E. Balamou (2005), *Rural/Urban Impacts of CAP Measures in Greece: An Inter-regional SAM Approach*, Working Document, Department of Economics, University of Patras.

Psaltopoulos, D., K.J. Thomson, S. Efstratoglou, J. Kola and A. Daouli (2004), "Regional SAMs for Structural Policy Analysis in Lagging EU Rural Regions", *European Review of Agricultural Economics*, Vol. 31, pp. 149-78.

Psaltopoulos, D. and K.J. Thomson (1998), *Methodology Working Paper for Impact Analysis (Task 2), FAIR3-CT 1554 Research Project on 'Structural Policies Effects on Poor, Remote Rural Areas Lagging behind in Development'*, University of Aberdeen.

Roberts, D. (1992), "UK Agriculture in the Wider Economy: An Analysis Using a Social Accounting Matrix", unpublished PhD thesis, University of Manchester.

————— (2003), "The Economic Base of Rural Areas: A SAM-based Analysis of the Western Isles, 1997", *Environment and Planning A*, Vol. 35, pp. 95-111.

Sarris, A. (ed.) (2003), *Towards a Development Strategy for Greek Agriculture*, Department of Economics, University of Athens.

Thomson, K.J. (2004), *Agricultural Multifunctionality and EU Policies: Some Cautious Remarks*, paper prepared for the ENARPRI Seminar, Andros, 2 October 2004.

Tsiboukas, K. (2003), *An Analysis of Impacts of Modulation and Decoupling in Greece*, Department of Agricultural Economics and Development, Agricultural University of Athens.

Van Huylenbroeck, G. and G. Durand (eds) (2004), *Multifunctional Agriculture: A New Paradigm for European Agriculture and Rural Development*, Aldershot: Ashgate.

11. Multifunctionality and the implications for EU policies

Yiorgos Alexopoulos, David Baldock, Dimitris Damianos and Janet Dwyer

Introduction

This chapter offers an overview of the final results of our quantitative analyses of different agricultural policy scenarios and their potential impact upon multifunctionality in EU agriculture. Our aim here is to review and discuss these results, to outline some implications for EU policies of the findings of ENARPRI partners' research work and, finally, to propose issues that call for further consideration and analysis.

11.1 Results of quantitative analysis: The multifunctionality perspective

11.1.1 Ireland

The Irish team used a FAPRI-Ireland model, i.e. a set of dynamic, multi-product, partial equilibrium commodity models, in order to estimate indicators such as animal numbers and input usage volumes. The model's estimates were incorporated into satellite greenhouse gas (GHG) models to make projections for multifunctionality indicators such as GHG and ammonia emissions. The researchers examined the 2003 reform of the common agricultural policy (CAP) with full decoupling as the baseline scenario and a 'most likely reform' scenario of the World Trade Organisation (WTO). The latter scenario involved the elimination of export subsidies over 10 years, an easing of market access by a 60% average tariff cut and an additional 70% cut in amber box domestic support (total aggregate measure of support).

The Irish study suggested that the number of dairy cows would decline under both scenarios with a slightly greater reduction occurring under the WTO reform (a 14% decline under the baseline and only 1% greater under the WTO reform scenario). Both scenarios also project decreases in beef cattle and sheep numbers up to 2015. Lower beef prices across Europe and Ireland, resulting from the expected fall in exports and rise in imports, are considered to cause the greater reduction in Irish cattle numbers projected under the WTO reform scenario. Livestock production is expected to become more extensive. The baseline scenario indicates shifts from areas cultivated with cereals and root crops into pasture. The relatively modest effects of the WTO reform scenario on sectors other than dairy are explained by reference to the more significant policy changes within the Luxembourg agreement that preceded the 2003 reform.

The reduction in overall agricultural activity, mainly in the dairy sector (which is the main source of agricultural GHG emissions in Ireland), will lead to a 14% cut in GHG emissions by 2015 relative to 2004 under the baseline scenario. Although WTO reform would deliver additional environmental benefits, further reductions in comparison with the baseline scenario appear to be only modest (only an additional 1% decline in ammonia emissions and 3.35% in total CO_2 equivalent by 2015 relative to 2004). Thus, a more radical WTO reform would be required in order to further reduce emissions from the dairy sector (Donnelan & Hanrahan, 2006).

11.1.2 Finland

The Finnish research team utilised the DREMFIA model – a dynamic, regional sector model that enables an analysis of Finnish agriculture. A technology diffusion model is combined with an optimisation routine that stimulates annual production decisions and price changes. The scenarios examined included the 2003 CAP reform as implemented in Finland, which means certain sectors remain partially coupled. The scenario of full decoupling was also investigated, along with the application of a simplified cross-compliance regime. The Finnish team also tested a scenario involving 20% cuts in CAP pillar I payments. The indicators that the model was able to examine included production levels, land use, nutrient balances, two types of biodiversity indexes, the use of pesticides, farm incomes and employment.

The Finnish study indicated that under the 2003 CAP reform, milk production in northern Finland would be reduced substantially and extensification in the sector would be expected. Beef production and grassland area would decline across the whole country. At the same time, the cropped area is predicted to decrease significantly as more land is put into 'green set-aside' (i.e. fallow) areas, particularly in the centre of the country. Crop area is expected to increase slightly in the north but shrink to some extent in the south. Pesticide use is expected to reduce in the short run as well as nitrogen and phosphorous surpluses. Beef and dairy production is expected to become more concentrated regionally, leading to notable rises in nutrient loading in some locations by 2015. The growth in fallow land will encourage improvement in habitat biodiversity in northern Finland. Nevertheless, the general national outlook for biodiversity as measured by the Shannon index is unfavourable, with the exception of northern Finland, because land-use diversity in most regions is expected to be reduced by the reforms. Cross compliance is predicted to enhance biodiversity conditions relative to the 2003 reform scenario, while reducing decoupled support by 20% apparently has little impact. Total farm incomes per farm are predicted to rise under all the scenarios, but total farm income will fall in the north as agriculture declines in this region. Farm employment will also fall slightly, under all the scenarios (Lehtonen, Lankoski & Niemi, 2005).

11.1.3 Czech Republic

In the Czech Republic, the research team used a non-linear optimising model (FARMA-4), which is designed to optimise the behaviour of three prevailing farm types, linked to certain multifunctionality indicators. The scenarios examined were the specific model of the CAP implemented in the country in 2002, full decoupling, full decoupling with decoupled direct payments reduced by 20% and, finally, the expansion of an agri-environment scheme across the country designed for grassland conservation. The indicators investigated were agricultural employment, production types, stock numbers, land use and fertiliser inputs.

The model suggested that full decoupling could lead to an extreme level of extensive farming with increased set-aside levels in the most productive regions. Reduced decoupled payments were expected to have a negative impact on the level of farm profitability. Increased less-favoured area payments counterbalance these trends in the regions. In any case agricultural output decreases and, in total, employment is expected to fall. The authors conclude that the appropriate implementation of pillar II measures may preserve the projected environmental benefits and counterbalance, to an extent, the negative consequences of reduced decoupled payments on rural employment (Doucha & Foltyn, 2006).

11.1.4 Greece

In Greece, the team examined policy effects by means of social accounting matrices at the national and regional levels. The region investigated was Archanes in the prefecture of Iraklion, Crete, a NUTS IV area that depends heavily on agriculture and specialises in the production of olive oil, table grapes and wine. At the same time, the region has a notable record for the implementation of pillar II policy measures. The models were constructed to assess changes in farm and non-farm sectors, in response to all four specified scenarios. The indicators analysed were farm output, farm and other employment, agricultural land use and levels of pollution emissions. The study incorporated an examination of the effects of CAP pillar II measures, which in Greece place most emphasis on the improvement of agricultural holdings' competitiveness. The models allow analysis of the effects of both pillar I and pillar II measures and policy instruments. The approach is static, however, and does not allow the dynamic assessment of changes in technology and innovation or the variables associated with structural adjustment.

Under a status quo scenario, at the national level the study suggested a significant decline in agricultural employment, the full impact of which was expected to exceed 10%. This result was related to the anticipated effect of decoupling on certain sectors in the agricultural economy, such as tobacco. Decoupling seems to be responsible for a projected reduction in total agricultural output by over 5%. In the case of tobacco, the fall in output was expected to approach 40%, whereas the respective reduction for the livestock sector was estimated to be slightly less than 2%. In general, the contraction of farm activity leads to environmental benefits. Yet, in contrast to what might be expected in most member states, these benefits are more than outweighed by rises in pollution emissions, which stem from the implementation of specific pillar II policy instruments associated with investment incentives for improving the sectors' competitive position. Pillar II activities generate positive outcomes for total employment (3%), farm employment (1.5%) and farm output (1.6%), which counterbalance the effects of decoupling in the agricultural sector.

At the regional level, under the status quo scenario a predominantly rural area such as Archanes sees agricultural output decline by almost 6% and rural employment by over 5%, mainly owing to a significant projected reduction of olive oil production by over 18%. When considerations of pillar II measures are incorporated, the effects on agricultural output and rural employment are subdued and become rather modest but still point in the same direction. As with the national-level analysis, pillar II measures are expected to cause an increase in pollution emissions at levels sufficient to minimise the environmental benefits of the contraction of farming activity.

Results for the scenarios reflecting further reductions in support have the same implications and the effects are intensified. Thus, the elimination of export subsidies leads to significant cuts in agricultural output and farm employment (both greater than 10%), whereas the impact of a further reduction in subsidies is rather marginal relative to the previous scenario. In these cases the increase in pillar II funds seems to have only marginal effects on agricultural output, agricultural employment and land abandonment.

In assessing the effects of pillar II measures as implemented in Greece to date, whatever scenario is used, one concludes that increased pollution emissions are generated. In certain cases such rises in pollution overwhelm the environmental benefits deriving from the contraction in farming activity. As pointed out by the authors, the question of whether pillar II measures as currently designed and implemented in Greece are best suited to agriculturally-dependent local economies that suffer from a severe reduction in pillar I-types of support, is pertinent (Psaltopoulos & Balamou, 2005).

11.2   Review and discussion of the results

11.2.1 Environmental indicators

Considering first the environmental indicators and their responses to the different policy scenarios, the predictions of the models indicate some positive and some negative repercussions of decoupling. Lower market prices, *ceteris paribus*, will tend to lead to extensification (especially in livestock sectors). This outcome would imply environmental benefits in terms of water, GHG emissions, biodiversity and landscapes in intensively farmed situations and regions. The same trend, however, would indicate greater environmental risks in extensive (marginal) habitats where the loss of stock could threaten the viability of traditional management systems and thus the implications for biodiversity and landscapes could be negative. Nevertheless, the models also suggest marked inter-regional shifts and that some effects are only temporary (e.g. with regard to input-use levels in Finland). In sum, the models point to a highly complex set of environmental outcomes in relation to the effects of decoupling – which is also the view of other experts in this area (e.g. GFA–RACE & IEEP, 2004).

11.2.2 Socio-economic indicators

On the socio-economic side, the models indicate some trends towards reduced levels of farm employment and lower farm incomes following a change in policy, although not in all cases. There are different implications of these changes depending on the capacity of other sectors to absorb labour. For example, in an otherwise buoyant rural economy shrinkage in farm employment can be positive because it signifies a redeployment of labour into sectors with higher productivity, whereas in an otherwise stagnant rural economy it can be a problem because labour lost from farming may simply increase unemployment. The story is different for the Czech Republic and probably other new member states as well because the phasing-in of pillar I support in these countries may stimulate an expansion in the farm sector, leading to both positive and negative consequences for sector employment, contingent on the precise nature of this expansion. This effect is likely to dwarf the impact of decoupling, given that in most of these countries governments have been operating a simplified pillar I payment system rather than the sector-specific direct payments under the main regimes. Furthermore, in those few countries that had implemented a more sector-specific system (including the Czech Republic), this experience will have been relatively short-lived.

11.2.3 Rural development indicators

When considering the multifunctionality effects of scenarios that include the expansion of pillar II-type supports alongside decoupling, the indicators suggest potentially positive results, but only if these measures are effectively targeted. It should be noted that relatively few teams were able to model this pattern and that in most cases, only simplified pillar II schemes were modelled. An expansion in agri-environment schemes (as modelled in Finland and the Czech Republic) generally implies positive environmental benefits, but in the case of Greece, where most of the current pillar II aid is devoted to enhancing agricultural competitiveness, an expanded pillar II suggests negative environmental outcomes. In both cases, where farm employment was examined, the results of expanded pillar II expenditure were negative for this indicator. Still, it is known that this outcome is highly conditional upon the precise design of pillar II measures and that some schemes, including agri-environment ones, have been shown to have positive effects upon farm employment (Dwyer & Kambites, 2005).

The findings of the modelling work in relation to pillar II policies are inconclusive, largely because these policies vary considerably among different areas. In addition, many are difficult to model in a precise way because the causal relationships between measures and outcomes are only partially understood and generally cannot be reduced to basic production, price and input responses, as with pillar I measures. Yet from the limited results of the relevant modelling exercises and from a wider literature review, the implications of the Doha round indicate that there is a need for policy-makers to seek to use the resources available for pillar II actions in a clearly targeted way in future. Pillar II activities could be more explicitly designed to promote the positive and minimise the negative repercussions predicted as a result of pillar I decoupling. Nevertheless, these findings have to be viewed in the context of current and perhaps more significant domestic issues as regards the future scale and direction of pillar II support in Europe. Of particular significance is the agreement on the EU budget for 2007–13 arrived at in December 2005. The agreement has severely reduced the budgeted funds available to this pillar for the next seven-year programming period in the EU-15 with a more positive result in the new member states.

11.2.4 Impact of the Doha round on multifunctionality

An increase in market access combined with lower market prices in the EU for many agricultural commodities would tend to suggest a contraction in the domestic farm sector – i.e. fewer farmers in the EU producing lower levels of agricultural output. At the same time, there are likely to be both gains and losses for the quality and diversity of the rural environment, the pattern of which will be highly dependent upon the farming systems currently in place in different regions and sub-regions of Europe. Changes in the rural environment will also depend on several broader factors such as rural demography and non-farming economic opportunities. In respect of the wider social implications of the trade round, it seems clear that there will be adjustments as a result of Doha that could lead to negative and positive effects at the local level. The balance of these will again be highly contingent upon the nature of the local economy in each particular region or sub-region. It also seems likely that the trade round will take place against a background of continuing and accelerating structural change in the new member states, which is largely the result of their accession to the EU and is much less related to overarching international factors.

The results of the modelling research carried out in each case study contribute to these observations, and indicate that the effects on several multifunctionality indicators are regionally dependent and differentiated among sectors and product mixes. Thus, the Irish case study demonstrates that the effects on agricultural activity, other than dairy, stem mainly from the CAP reform associated with the Luxembourg agreement while the impact of the hypothetical Doha reform would be rather modest. Along the same lines the full-decoupling scenario is likely to have a greater impact on atmospheric emissions than the projected Doha reform – although assumptions about market access are important here. In the Greek regional case study, following a quite different methodology, the estimated negative trends are attributed equally to decoupling and the expected decline in prices. In-depth analysis, however, suggests that specific sectors are not equally affected by policy measures of a similar kind. More specifically, in the case of the two products of most importance for the local agricultural economy in the Crete case study, namely olives and vines, it is argued that while decoupling affects primarily the olive oil sub-sector it is the reduction in prices that mostly impacts the returns and viability of the table- and wine-grape producers. Yet, price reductions also generate a greater amount of environmental benefit than decoupling.

11.3   Implications of model findings

Heterogeneity of the models, scenarios and situations

In an effort to capture, explore and assess the key effects of the 2003 reform from the point of view of multifunctionality, a group of institutes managed to specify and test several models. These models, briefly outlined and commented upon above, were constructed under various assumptions relevant to the member states and regions of application. Additionally, by means of the same models, researchers attempted to investigate some further reform possibilities as indicated by a range of scenarios, each one of which represented a different, alternative reform outcome. Owing to practical difficulties stemming from a lack of homogeneity among different country situations, the national models were set up in a way that did not allow the testing of an identical set of scenarios for all the member states involved. Significant variations, even in the baseline scenario, were unavoidable as the 2003 agreement and the subsequent decoupling adjustments triggered by the WTO were implemented in a variety of ways in the member states in advance of Doha.

Doha's 'modest impact'

An interesting point about the models' results is that, generally, trade-driven policy changes in the EU and the member states of the kind assumed here appear to have a limited impact upon agriculture as well as selected multifunctionality indicators. This general observation has to be treated with caution, however, as the following specific factors need to be considered:

- As the 2003 CAP reform is often justified on the basis that it was designed to assist negotiations under the Doha round, it would be possible to recognise it as being an important consequence of the WTO process even though the CAP reform was decided upon prior to the round's conclusion. In this context, the network's policy scenarios for Doha outcomes were seen and designed as incremental adjustments to the direction imposed by the basic model implemented in 2003. Thus, they correspond to only mild incremental changes in respect of some of the multifunctionality effects that the reform is predicted to have. In general, the implementation of the 2003 CAP reform was forecasted by several models to have quite significant effects upon agriculture and thus upon various aspects of multifunctionality. These effects were largely attributable to the combined consequences of decoupling and continued decreases of guaranteed prices with only limited compensation in the major supported sectors.

- The modelling exercises for Finland and Greece in particular showed that, as expected, the effects of similar Doha scenarios at the EU or member state level would not necessarily give rise to equivalent effects at the regional and local levels. At these latter two levels, the results may be more significant for specific multifunctionality indicators (e.g. biodiversity, concentration of production, nutrient loadings, land taken out of production, extensification and agricultural employment). The findings indicate that because of the spatial heterogeneity of the socio-structural characteristics of agriculture and rural areas throughout Europe, the examination of likely sub-regional and regional outcomes of future policy changes (although challenging) will be important.

- The results of the dynamic modelling of outcomes in the Finnish case study suggest that some effects can change substantially over time. This factor should be taken into account when studying policy effects, such as those under the Doha scenarios, on certain multifunctionality indicators. The examples of fertiliser surpluses and pesticide use, which were predicted by the model applied in Finland to decrease in the short run but increase again at a later point in time as farms adjust and restructure in response to policy changes, are important here.

- In the new member states, the 2003 reform scenarios are set against an underlying trend in policy that is quite different to that which operates in the EU-15. Levels of domestic support have been significantly increasing over time as common pillar I supports are phased in. Farmers are in essence being offered much larger payments, which, even when decoupled are likely to give rise to a greater capacity to produce and raise profits. This result is attained by means of the phasing-in of pillar I support over the period for which the network's models were run. The results of the research work carried out for Czech agriculture demonstrate that an increased pillar I-type of support favours larger and more commercial farm enterprises at the expense of the rest. When combined with full decoupling, commercial farms are expected to only utilise land of better quality whereas other farms with semi-subsistence production will probably keep all their land in production, irrespective of quality.

- It can be seen that some variables, such as farm labour, farm size and concentration are somewhat 'insulated' with regard to the external policy environment, owing to their structural nature. Thus they usually display distinctive long-term trends and show significant responses to policy changes.

- Some variables are particularly influenced by local or regional conditions and peculiarities, such as the prevailing political choices and preferences; hence, they are not necessarily reflective of wider policy changes and become very difficult to model. Member states' rural development programmes, which are characterised by a highly variable mix of measures, or the different approaches that governments adopt in order to cope with the problems in their less-favoured areas, are good examples that explain important differences in environmental outcomes among countries.

- Institutional factors such as the national or regional government's role as regulators in the agricultural sector or the rural economy as a whole are very decisive in determining the final outcome. For instance, governments at different levels intervene in the sphere of land-use policy; they also design and implement land tax systems as well as monitor, supervise and control CAP pillar I and pillar II measures, thus influencing farmers' decisions to abandon or reclaim land. Such issues can have a great impact on most multifunctionality indicators at the ground level.

Pillar II rural development measures

A limited number of models can take into account pillar II rural development policies and their impact. Examples include the social accounting matrices used in the case of Greece, the Finnish model and the optimisation (FARMA-4) model applied in the Czech Republic. For this type of modelling a considerable range of assumptions is unavoidable, especially when looking ahead at prospective programmes that have not yet been agreed. Such assumptions include estimates of the measures that member states will adopt in 2007–13, the levels of take-up by farmers and others, the likely outcomes and the ramifications they will have. The case studies summarised here have demonstrated (as expected) that member states vary greatly in the approaches they have adopted and, consequently, multifunctional variables and indicators cannot be expected to follow a uniform pattern.

The Greece case study is extreme since the model depicts the potential consequences of rural development measures that are substantially focused in budgeting terms on the improvement of agricultural competitiveness. It comes as no surprise that a switch to this form of rural development support is expected to have an unfavourable impact on multifunctionality indicators. In the very different case of the Czech Republic, where pillar II measures place clear emphasis on agri-environment issues (namely, the expansion of a scheme designed for grassland conservation), the respective model concludes that the effects of these rural policy instruments are distinctly positive for the environment. There will be many examples of programmes within this range with varying outcomes. Although there are ceilings on the percentage of funds that can be allocated to the main categories (cuts to measures under the new rural development regulations for 2007–13), major differences will persist, creating a considerable challenge for both forecasting measures and modelling policy scenarios. Rural development measures per se are not invariably associated with the enhanced provision of goods, as illustrated in Greece.

Cross compliance

If implemented according to guidelines and obligations commonly agreed upon by the member states and the European Commission, cross compliance should be expected to have positive repercussions on a number of multifunctionality indicators such as the levels of agricultural pollution, nature management and the control of land abandonment. These benefits could help to address some of the negative effects of decoupling, such as marginalisation and scrub invasions, which might otherwise occur according to theory and literature (see above). Although cross compliance has not been treated by the models in quantitative terms, certain interesting issues are being raised. A number of the research teams responsible for the implementation of case studies have pointed out that 'good farming practices' associated with eligibility for pillar II payments are not effectively enforced in the respective member states. In general the level and manner of implementation throughout the EU is uncertain at this point in time. In a few years, however, empirical evidence of how cross compliance has worked will be available and could well be employed in quantitative and qualitative policy-impact analyses.

Some considerations about model effectiveness

Models, in general, are not equally convincing and forceful when addressing different parameters. Models are better suited to the treatment of simple, quantifiable parameters such as employment, farm structures, certain forms of input use and outputs (including pollution) even if their relationship to policy variables is quite complex. Models are less effective when they are called upon to handle less straightforward multifunctionality indicators such as landscape, cultural factors and biodiversity. Furthermore, in assessing the multifunctionality impact of agricultural policy reform in a holistic way, models should be designed and constructed to investigate links in the food supply chain. It is also important that the models assess how the impact of policy changes in Europe could affect certain countries outside the EU (e.g. the effects of the Doha round on environments and employment in developing countries).

The discussion in this section leads us to another conclusion, namely that the assessment of policies and their consequences in relation to multifunctionality cannot be undertaken entirely through simple economic welfare analyses and models solely tailored to suit this perspective. Additional information on diversified socio-structural characteristics among member states and regions, farmers' and consumers' attitudes and preferences along with the role of institutions needs to be captured and incorporated into the analyses as far as possible.

11.4   Issues that call for consideration and analysis

We draw a number of conclusions from this exercise so far, as follows.

- Current models that seek to predict the response of agriculture in particular countries to likely changes in policy as a result of the Doha round can examine the ramifications on a broad range of potential indicators of multifunctionality.

- In general, the majority of predicted results in relation to multifunctionality indicators are plausible from the point of view of informed expert understanding, as measured by the forecasted direction and scale of change. Yet in some instances there can be unexpected anomalies that appear to stem from shortfalls in the modelling process rather than real-life challenges to the expert views. The models tend to adopt rather simplistic, standardised relations between farm sector change and the variety of multifunctionality indicators and, as with most modelling of this kind, the predicted outcomes are very dependent on the assumptions made in defining these relations.

- The models tend to pick up simple functional relationships – for example where certain environmental outcomes can be portrayed as negative externalities such that more farming usually means less environment in a simple trade-off relationship – but they do less well where the effects are more complex. They are also more able to deal with linear relationships that can be objectively defined than with context-sensitive relationships where the perception of the effects can be contested. Also, national-level modelling approaches have limited ability to pick up spatially differentiated responses, whereas the regional ones have a tendency to suggest that regional and even sub-regional variability of responses will be highly significant for multifunctionality.

- Because the models that we have examined here are for the most part nationally or regionally defined and are based on different approaches and methodologies, their results are not strictly comparable across member states or regions. Nevertheless, some comparisons may be valid in the context of expert interpretation of their overall findings.

- Some of the proposed multifunctionality indicators are potentially ambiguous. For example, a contraction of the farm workforce that comes as a result of policy reform can be seen as a positive development in cases where former farmers and their families diversify their income sources by spending more outside conventional farming. On the other hand, in a less buoyant economy, a reduction of employment in agriculture may lead to unemployment owing to a real or perceived lack of alternative sources of jobs or income. In particular areas this could result in outward migration and damage to traditional rural culture.

Standing back from our modelling studies, it is possible to consider several important issues in relation to the multifunctionality/trade debate.

There are strong arguments for incorporating multifunctionality objectives more explicitly in agricultural policy, not least in relation to changes in trade policy. The 2003 reform of the CAP was intended as a step in this direction, whatever the eventual outcome. In general, conservation and environmental quality provide a transparent and publicly acceptable basis for rural household support, which is also consistent with policy reforms that are acceptable in the framework of the WTO. Yet the efficient implementation of environmental policy in agriculture presupposes that support is addressed to agricultural activities that provide particular benefits to the natural environment. A policy favouring the adoption of a clear environmental basis for support would probably imply a redistribution of payments that is significantly different from that which currently prevails under the CAP. If farm payments are to be justified on the basis of a wide spectrum of environmental goals and multifunctionality indicators (biodiversity, landscape, historical features, water, soil and air quality), then a significant proportion of agricultural holdings might qualify for assistance at some level (Smith, 2001). In this respect, the critical issue is perhaps where the 'reference level' of responsible farming practice lies, below which no payment is justified whereas above it farmers should be rewarded for providing these public services.

CAP decoupled transfers, along with a host of other income sources contribute to total household income. On the basis of a number of socio-economic factors (age, preferences, wealth, family characteristics, location, the role of institutions, etc.) every farm household decides how to allocate income between consumption and savings/investment. In principle, households compare possible alternative rates of return in order to decide about how to invest. They keep on investing in the farm until expected returns cease to exceed returns associated with off-farm opportunities. In theory, decoupling alone should have practically no effect on farm investment; neither should it affect production levels. Nevertheless, decoupling will in practice influence production decisions when market failures exist, such as credit constraints that prevent farmers from profiting from farm investments. In such cases, decoupling might make it possible for the household to proceed with an investment that, in turn, might lead to a small increase of production (Burfisher & Hopkins, 2004). In practical terms, no agricultural policy tool designed to support producers' incomes appears to be entirely production-neutral (OECD, 2001).

A multifunctionality-oriented policy promoting environmental benefits from extensive agriculture needs to be related to the risk-averse behaviour of farmers. As variability in the market for agricultural goods and thus the level of market risk increases, as could be anticipated after the abolition of coupled support and adoption of the single farm payment (SFP) scheme, farmers can be expected to be less inclined to adopt more intensive practices (Oglethorpe, 1995). This consequence could bring about some 'environmental dividend' in certain intensively farmed areas of the EU. At the same time, however, decoupling could decrease the incentive to farm in marginal areas including some of high nature value that are dependent upon the continuation of extensive farming systems, in which case negative environmental benefits could result. Hence decoupling, as a policy option, is clearly insufficient on its own to provide a socially optimal level of environmental benefit from agriculture.

From the strict perspective of economic theory, there should be a clear distinction between policies designed to enhance farmers' incomes and welfare and policies intended to promote multifunctionality in agriculture. Policies aiming at income enhancement have a focus on the farmers' consumption abilities, whereas policies where multifunctionality is the primary goal are focused on farmers as producers of private and public goods (Prestegard, 2003). Yet in reality, it is often the case that instruments designed for the former purpose have been reinterpreted in the latter context; thus conversely, agri-environment and other pillar II payments have been perceived as providing an element of income support in many areas of Europe.

It is possible that decoupled SFPs will allow marginally viable, small agricultural holdings to remain in operation for a longer period than would otherwise be the case. This category of farm might be able to cover variable expenses and thus remain in business in the short run. It is highly unlikely however, that large numbers of scarcely viable farms will remain in production in the long run because of rising land values and increasing competition. SFPs could offer a financial cushion enabling them to maintain their low-yield production and thus prevent structural change. By contrast, SFPs could be used by larger producers to buy production rights along with the land that is required, thus triggering greater structural change than might otherwise have occurred. The more efficient producers are likely to adopt newer technologies and exercise better management techniques, potentially raising yields and production levels. Therefore, the direction of the net effect of SFPs on restriction and consolidation is uncertain (Burfisher & Hopkins, 2004).

The long-term decline in the farm labour force is driven by the 'pull' process of work off the farm and the 'push' process attributed to the adoption of labour-saving technology and declining returns to domestic producers as a result of increased market competition. It seems very likely that decoupled support, in the form of SFPs, will reinforce existing trends in this respect, thus leading to a further reduction of total work hours in agriculture. The introduction of decoupled support will probably leave use of labour by larger commercial farms fairly unaffected. Similarly, the smallest farms that are already highly pluriactive, for whom agricultural income is now a minor share of total household income, seem unlikely to shed significantly more labour as a result of SFPs. It seems most probable that labour losses will be concentrated in those middle-sized holdings where there is still scope to make cost savings through the adoption of less labour-intensive farming systems or through economies of scale gained through farm enlargement (Linares, 2003),

11.5   Final recommendations

Notwithstanding these points, it would be useful to develop more sensitivity to these issues in future trade modelling/policy analysis work, so that the EU's international negotiators can draw upon these resources in considering options and negotiating details of agreements.

This view suggests a need to improve multifunctionality models at the country or regional level so that their predictions can more fully inform wider debates. At least three areas of improvement would seem particularly worthwhile:

1.   the introduction of spatial variation into the national models so that the differential effects on regions and sub-regions can be explored;

2.   an attempt to include dynamic representations of effects upon other sectors beyond the farm gate, in examining the wider socio-economic aspects of multifunctionality; and

3.   the identification of better means to incorporate the effects of CAP pillar II policies (as these become a more significant feature in EU countries) on sectoral and wider social and environmental para-meters, if multifunctionality issues are to be adequately examined.

At this stage, the models raise questions about the effects of trade-driven policy changes including both the political benefits and downsides of decoupling. They provide a useful complement to more qualitative judgements. Yet they also point to the need for more effective use of a range of analytical tools and data sources in examining and explaining complex phenomena in the rural economy and society, including the results of policy evolution.

Nevertheless, it seems clear that the modelling work that has been undertaken through the ENARPRI network has added to our understanding of the potential indications of multilateral trade agreements on the multifunctional character and qualities of agriculture across the EU, in different national and local contexts.

Bibliography

Burfisher, M.E. and J. Hopkins (eds) (2003), *Farm Payments: Decoupled Payments Increase Households' Well-being, not Production*, Economic Research Service, Amber Waves, USDA, Washington, D.C., February.

——————— (2004), *Decoupled Payments in a Changing Policy Setting*, Agricultural Economic Report No. 838, USDA, Washington, D.C., November.

Donnelan, T. and K. Hanrahan (2006), *Potential WTO Trade Reform: Multifunctionality Impacts for Ireland*, ENARPRI Working Paper No. 16, CEPS, Brussels, May.

Doucha, T. and I. Foltyn (2006), *Modelling the Multifunctionality of Czech Agriculture*, ENARPRI Working Paper No. 17, CEPS, Brussels, May.

Dwyer, J. and C. Kambites (2005), *Evaluation of agri-environment schemes in the EU: National report for the UK*, Study for the European Commission, DG Agriculture, led by Oreade-Breche consultants, France, European Commission, Brussels.

GFA–RACE and IEEP (2004), *The environmental implications of the 2003 CAP reforms in England*, Report to the Department of Environment, Food and Rural Affairs, London (retrieved from www.defra.gov.uk).

Guyomard, H. and K. Le Bris (2004), *Multifunctionality, Agricultural Trade and WTO Negotiations: A Review of Interactions and Issues*, ENARPRI Working Paper No. 4, CEPS, Brussels, December.

——————— (2004), *Multilateral Agricultural Negotiations and Multifunctionality: Some Research Issues*, ENARPRI Policy Brief No. 4, CEPS, Brussels, February.

Lehtonen, H., J. Lankoski and J. Niemi (2005), *Evaluating the Impact of Alternative Policy Scenarios on Multifunctionality: A Case Study of Finland*, ENARPRI Working Paper No. 13, CEPS, Brussels, July.

Linares, D. (2003), "Structure of Agricultural Holdings in the EU: Part-time Work on Agricultural Holdings", Statistics in Focus, Agriculture and Fisheries, Theme 5-29, Eurostat, European Commission, Brussels.

OECD (2001), *Multifunctionality: Towards an Analytical Framework, Executive Summary*, OECD, Paris.

Oglethorp, D.R. (1995), "Sensitivity of farm plans under risk averse behaviour: A note on the environmental implications", *Journal of Agricultural Economics*, Vol. 42, No. 2.

Prestegard, S.S. (2003), *Policy Measures to Enhance a Multifunctional Agriculture: Applications to the WTO Negotiations on Agriculture*, Norwegian Agricultural Economics Research Unit, Oslo.

Psaltopoulos, D. and E. Balamou (2005), *Modelling the Effects of Trade Policy Scenarios on Multifunctionality in Greek agriculture: A Social Accounting Matrix Approach*, ENARPRI Working Paper No. 14, CEPS, Brussels, September.

Smith, K. J. (2001), "Retooling Farm Policy", *Issues in Science and Technology*, National Academy, University of Texas, Dallas, Summer.

Glossary of Abbreviations

ACP	African Caribbean and Pacific countries
AMAD	Agricultural Market Access Database
AMS	Aggregate measure of support
AoA	Agreement on Agriculture
AVE	*Ad valorem* tariff equivalent
CAP	Common agricultural policy
CCPs	Counter-cyclical payments
CEECs	Central and Eastern European countries
CGE	Computable, general equilibrium
DDA	Doha Development Agenda
DREMFIA	A partial equilibrium recursive model used in the case study of Finland
EAGGF	European Agricultural Guidance and Guarantee Fund
EBA	Everything but Arms
EEC	European Economic Community
EPAs	Economic Partnership Agreements
FAO	Food and Agriculture Organisation
FADN	Farm Accountancy Data Network
FAPRI	An economic, partial-equilibrium, agricultural commodity and inputs model used for the case study of Ireland
FARMA-4	A non-linear optimising model used in the case study of the Czech Republic
FTA	Free trade agreement
GAFTA	Greater Arab Free Trade Area
GATS	General Agreement on Trade in Services
GATT	General Agreement on Tariffs and Trade

GHGs	Greenhouse gases
GRIT	Generation of regional input–output tables
GSP	Generalised System of Preferences
GTAP	Global Trade Analysis Project
IO	Input–output
IPCC	Intergovernmental Panel on Climate Change
LDCs	Least-developed countries
LFAs	Less-favoured areas
MAcMap	Market Access Map, a database of customs tariffs or import duties of 178 countries
MEDA	Financial instrument of the Euro-Mediterranean Partnership
MERCOSUR	Customs union among Central and South American countries
MFN	Most-favoured nation
MPS	Market price support
MTR	Mid-term review of the EU's common agricultural policy
NCCS	National Climate Change Strategy for Ireland
NSSG	National Statistical Service of Greece
NUTS	NUTS is the EU nomenclature for territorial units for statistical purposes
OCTs	Overseas countries and territories
OECD	Organisation for Economic Cooperation and Development
PSEs	Producer support estimates
PTAs	Preferential trade agreements
ROW	Rest of the world
SAM	Social accounting matrix
SAPS	Simplified Administrative Payment Scheme

SDT	Special and differential treatment
SHDI	Shannon's diversity index
SPS	Sanitary and phytosanitary measures
SSG	Special safeguard mechanism
STEs	State trading enterprises
TAs	Trade agreements
TBTs	Technical barriers to trade
TRAINS	Trade Analysis and Information System, an UNCTAD database
TRIPS	Agreement on Trade-Related Aspects of Intellectual Property Rights
TRQs	Tariff-rate quotas
UAA	Utilised agricultural area
UNCTAD	United Nations Conference on Trade and Development
URAA	Uruguay Round Agreement on Agriculture
VERs	Voluntary export restrictions
WTO	World Trade Organisation

Contributors

Yiorgos Alexopoulos is a Researcher at the Department of Agricultural Economics and Rural Development, Agricultural University of Athens.

Eudokia Balamou is a PhD candidate at the Department of Economics, University of Patras.

David Baldock is a Director at the Institute for European Environmental Policies (IEEP), London.

Martina Brockmeier is a Director at the Federal Agricultural Research Centre (FAL), Institute of Market Analysis and Agricultural Trade Policy, Braunschweig.

Dimitris Damianos is a Professor at the Department of Agricultural Economics and Rural Development, Agricultural University of Athens.

Crescenzo dell'Aquila is the Staff Chief for the Agriculture Branch at the Regional Ministry for Agriculture, Industry and Trade, Regional Government of Campania, Naples and a former Researcher at the Istituto Nazionale di Economia Agraria (INEA), Rome.

Trevor Donnellan is a Senior Research Officer at the Rural Economy Research Centre, Teagasc, the Irish Agricultural and Food Development Authority, Athenry, Galway.

Tomáš Doucha is the Deputy Director for Research at the Research Institute of Agricultural Economics (VUZE), Prague and the Deputy Minister for Agriculture of the Czech Republic.

Janet Dwyer is a Senior Associate and Reader in Rural Studies at the Institute for European Environmental Policy (IEEP) and the University of Gloucestershire.

Ivan Foltýn is a Researcher at the Research Institute of Agricultural Economics (VUZE), Prague.

Hervé Guyomard is the Director of Research at the Institut National de la Recherche Agronomique (INRA) and a Professor at the Centre d'Etudes Prospectives et d'Informations Internationales (CEPII), Paris.

Kevin Hanrahan is a Researcher at the Rural Economy Research Centre, Teagasc, the Irish Agricultural and Food Development Authority, Athenry, Galway.

Ellen Huan-Niemi is a Principal Research Scientist at MTT Agrifood Research Finland, Helsinki.

Hans Jensen is a Research Fellow at Institute of Food and Resource Economics (FOI), the Royal Veterinary and Agricultural University, Copenhagen.

Eleni Kaditi is a Research Fellow at the Centre for European Policy Studies (CEPS), Brussels.

Rainer Klepper is a Researcher at the Federal Agricultural Research Centre (FAL), Institute of Market Analysis and Agricultural Trade Policy, Braunschweig.

Marijke Kuiper is a Researcher at the Agricultural Economics Research Institute (LEI/Wageningen UR), The Hague.

Marianne Kurzweil is a Researcher at the Federal Agricultural Research Center (FAL), Institute of Market Analysis and Agricultural Trade Policy, Braunschweig.

Jussi Lankoski is currently working at the OECD Directorate for Food, Agriculture and Fisheries, Paris.

Heikki Lehtonen is a Principle Research Scientist at MTT Agrifood Research Finland, Helsinki.

Jyrki Niemi is a Professor in Agricultural Policy at MTT Agrifood Research Finland, Helsinki.

Janine Pelikan is a Researcher at the Federal Agricultural Research Centre (FAL), Institute of Market Analysis and Agricultural Trade Policy, Braunschweig.

Demetrios Psaltopoulos is an Assistant Professor at the Department of Economics, University of Patras.

Petra Salamon is a Researcher at the Federal Agricultural Research Centre (FAL), Institute of Market Analysis and Agricultural Trade Policy, Braunschweig.

Johan Swinnen is the ENARPRI Coordinator, a Senior Research Fellow at the Centre for European Policy Studies (CEPS) and a Professor at the Katholieke Universiteit Leuven.

Frank van Tongeren is a Senior Economist at the Directorate for Food, Agriculture and Fisheries, OECD, Paris.

Oliver von Ledebur is a Researcher at the Federal Agricultural Research Centre (FAL), Institute of Market Analysis and Agricultural Trade Policy, Braunschweig.

Wusheng Yu is an Associate Professor at the Institute of Food and Resource Economics (FOI), the Royal Veterinary and Agricultural University, Copenhagen.

ENARPRI Working Papers